3/2012

A Mathematical Medley

Fifty Easy Pieces on Mathematics

A Mathematical Medley

Fifty Easy Pieces on Mathematics

George G. Szpiro

Translated by Eva Burke

AMERICAN MATHEMATICAL SOCIETY

Some of these stories were published previously in the following German-language publications: "Mathematik für Sonntagnachmittag: Weitere 50 Geschichten aus Mathematik und Wissenschaft", Verlag Neue Zürcher Zeitung, Zürich, 2006, and "Mathematischer Cocktail: Zauberwürfel, die Liebe zu den Warteschlangen und weitere Geschichten", Verlag Neue Zürcher Zeitung, Zürich, 2008.

An earlier version of "Bella Abramovna Subbotovskaya and the "Jewish People's University"" was previously published in *Notices Amer. Math. Society* **54**, no. 10 (2007), pp. 1326–1330.

An earlier version of "Interview with Stephen Smale" was previously published in *Notices Amer. Math. Society* **54**, no. 8 (2007), 995–997.

2000 *Mathematics Subject Classification.* Primary 00Axx, 01Axx, 97–XX.

For additional information and updates on this book, visit
www.ams.org/bookpages/mbk-73

Library of Congress Cataloging-in-Publication Data

Szpiro, George, 1950–
A mathematical medley : fifty easy pieces on mathematics / George G. Szpiro ; translated by Eva Burke.
 p. cm.
Translation of 20 stories from Mathematischer Cocktail and 21 stories from Mathematik für Sonntagnachmittag, with nine new stories.
ISBN 978-0-8218-4928-6 (alk. paper)
 1. Mathematics—Miscellanea. I. Title.
QA93.S973 2010
510—dc22
                                                                                                    2009053215

Dedicated to the memory of our father
Symcha Binem Szpiro,
August 4, 1914–October 10, 2009

Contents

Contents

Choosing and dividing

Money, and making it

Interdisciplinary matters

Chapter 1

A Baker's Dozen

It is a curious truism that wherever one goes, one bumps into the number twelve: the tribes of Israel, the apostles of Jesus, the signs of the zodiac. One would therefore not be amiss in assuming that twelve and all things connected to this number must necessarily be good. Add one and the resulting number thirteen cannot but spell disaster as it intrudes on the round and fulsome dozen. The number seven, found in the colors of the rainbow, the days of the week, the number of continents, and the notes in an octave, obviously indicates unity and perfection. Hence, the number six must be imperfection. Repeat the six three times to form the number 666 and, woe is us, one has the number of the devil. Any nonsense can be proven with numerology. Nevertheless, number mysticism is not utterly senseless. As we shall see, it was the precursor to rational science.

Numerologists never tire of interpreting numbers, describing attributes and prophesying events which can be connected in some way to numbers. Even if this seems an outlandish pastime, more suitable to those who dabble in the occult, numerologists keep at it. It is a great comfort to them that most negative prognoses can be turned into their exact opposite—with an equal amount of certainty— by swiftly multiplying, dividing or simply reinterpreting the ominous number.

Mathematicians confronted with such humbug wrinkle their noses in disdain. Of course, they recognize that twelve is an important number but its preeminence is to be attributed not so much to its mystical properties than to the fact that a dozen can be divided without a remainder by 2, 3, 4, and 6, not to mention 1 and twelve itself. Twelve, therefore, has twice as many divisors as 10 which can be divided only by 2 and 5 aside from 1 and itself. This is the reason why twelve came to serve as the basis of the duodecimal system which prevailed in the Anglo-Saxon world. The Ancient Romans preferred the number ten which has the important advantage that school children and less numerically able business folks can count up to that number by using both of their hands.

Toward the end of the eighteenth century, the mathematicians Charles de Borda, Joseph-Louis Lagrange, and Antoine-Laurent Lavoisier fully recognized this advantage. They sided with the finger-counters and suggested to the French Academy that the decimal system be used as the only legal standard for measuring lengths and weights. (Borda further argued that the day should be divided into 10 hours with an hour divided into 100 minutes, each of 100 seconds, but this proposal never went very far.)

Let us turn to the number six. In antiquity, six was considered a perfect number because the sum of its divisors, other than itself, is equal to itself $(1 + 2 + 3 = 6)$. How about seven and thirteen? As far as mathematicians are concerned they are neither better nor worse than six or twelve, but they are of greater interest since they have no divisors, other than one and themselves. Such numbers are known as primes; these are the atoms which make up all the rest of the numbers.

Numerologists and other mystics who believe in the magical properties of numbers usually credit Pythagoras as their guide. The Greek philosopher did indeed try to understand the cosmos with the help of whole numbers and geometric shapes. While we might be tempted to consider many of his so-called discoveries simplistic, Pythagoras was, in fact, a pioneer and his dictum "all is number" was revolutionary. But the Pythagoreans' concept of the world remained severely limited. It was confined to natural numbers and fractions. As soon as his students realized that the diagonal of a square cannot be expressed as the ratio of two integers, the Pythagorean brotherhood's world view collapsed. Legend has it that the discoverer of the irrational numbers was put to death.

Attempts to understand nature and the cosmos with the help of numbers continued with Plato and after him with the Neoplatonists. In the third century A.D., the philosopher Iamblichus developed Neoplatonism into a "Theology of Arithmetic." In his work, Iamblichus alternates between Pythagorean concepts and free associations; numbers assume mystical symbolism: numerology was born.

At around the same time, Jewish Mysticism, known as Kabbala, gained prominence. The *Sefer Yetzirah* (Book of Creation), the oldest and most mysterious of all Kabbalistic texts, was written between the third and sixth century. The creation and the order of the universe are interpreted with the help of the numbers 1 to 10 and the twenty-two letters in the Hebrew alphabet. One is God, two is divine wisdom, three is worldly understanding. Love, power, and beauty follow, and so on. The second book of the Kabbala, said to have been written in around the thirteenth century, is the *Zohar*, (Book of Radiance) which too had a powerful impact on Jewish mysticism. One tool used in the Kabbala, although not taken seriously by most rabbis, is Gematria, from the Greek work geometry. By assigning numerical values to the letters of the Hebrew alphabet, the technique allows the calculation of numerical equivalences of letters, words, and phrases. Once a text has thus been reduced to its numerical value, this same numerical value can be re-expanded into different words and phrases. Thus, Gematria opens up myriad possibilities to interpret and prophesy, thereby exploring relationships between words and ideas.

It is surprising that scientists and theologians have always intuitively felt that it would be numbers, and only numbers, which can adequately describe the world. "True understanding of the divine is out of reach for those ignorant of mathematics," the German Cardinal Nicholas von Kues wrote in the fifteenth century. Today, of course, intuition has turned into conviction and we know that mathematics is the basis for our understanding of nature. Like the numerologists of old, modern natural scientists continue to ponder observations, in attempts to establish correlations between different sets of data.

It never ceases to surprise natural scientists that mathematics has remained the daily bread and butter of scientists. In a widely quoted essay, Eugene Wigner, winner of the Nobel Prize for Physics in 1963, spoke of "the unreasonable effectiveness of mathematics in the natural sciences." And Albert Einstein, throwing the Pythagorean world view definitely overboard, asked "How can it be that mathematics, being after all a product of human thought which is independent of experience, is so admirably appropriate to the objects of reality?" For him, the most incomprehensible aspect about the world was that it is comprehensible at all.

For mystical numerologists, in comparison, the going was easy. Everything which appeared to be plausible—and there is a lot that seems plausible to people of faith or superstition—was deemed legitimate. Where these mystical numerologists fell short was in the scientific method. They disregarded the necessity to confirm or refute a theory by rigorous experiments.

Galileo (1564–1642) was one of the first natural philosophers who refused to accept explanations of natural phenomena based merely on their plausibility, on theological revelations, or on the arguments of earlier authorities. He demanded that they be proven through experiments, observations, and rational reasoning. Nature, he wrote, is a book written in the language of mathematics. Today Galileo's approach is considered the only valid road to the understanding of the world around us, but in the sixteenth century this was definitely the exception.

One of Galileo's contemporaries who did accept the necessity of observations and the prevalence of mathematics, but who nevertheless remained caught up in mysticism and astrology, was Johannes Kepler from Prague. In 1594, then a 23-year-old theology graduate from the University of Tuebingen, Kepler embarked on his career as astronomer by investigating the movement of the planets known at the time: Mercury, Venus, Earth, Mars, Jupiter, and Saturn. His aim was to bring numerical order to their orbits. To the young Kepler this was of immense importance since he firmly believed in astrology.

Throughout his life, he would remain convinced that stars were imbued with magical powers. So, just like a student trying to make sense out of a series of numbers in an IQ test, Kepler sought to detect regularities in the data he was provided with. He added, subtracted, multiplied, and divided numbers to, from, with, and by each other; he used factors and constants and postulated invisible planets. It was to no avail; all his efforts were in vain. "I lost a lot of time playing around with these numbers," he would later state regretfully.

The epiphany occurred in 1595. Kepler, who had in the meantime advanced to the position of school teacher, was drawing a geometric figure on the blackboard when he suddenly had a brain wave: the planets' orbits ran along spheres that circumscribed interleaved Platonic solids. Checking his flash of enlightenment with sober calculations, Kepler saw his intuition confirmed. Remarkably, the margin of error was less than 10 percent and this lay within the accuracy of then available astronomical observations. A year later he published his insight in *Mysterium Cosmographicum* (Cosmic Mystery), a book that was greeted with enthusiasm by the professional world. After all, the harmonious interplay of the heavenly bodies was a brilliant confirmation of the Pythagorean world view. There was just one minor problem: his insight was totally and utterly false.

The moment of truth came several years later. One of Kepler's bitter rivals, the imperial mathematician to Emperor Rudolph II, Tycho Brahe, took issue with Kepler's interpretation. But without access to Brahe's significantly more accurate data, Kepler had no way of addressing the issue. Only after Brahe died and Kepler was appointed his successor, did he gain access to the observations. Then, at long last, Kepler was in a position to analyze Brahe's planetary observations and to complete his own tables. Finally he realized that the orbits of the planets are not perfect circles, but ellipses. Hence, they could not run around spheres. In Kepler's favor it must be said that he was honest and brave enough to own up to his previous mistake. In 1609 and 1619 he published *A New Astronomy* and *Harmony of the Worlds* in which he proposed three theses. These, for once, were correct and would henceforth carry his name: Kepler's Laws.

In his third law, Kepler drew a parallel between the time it takes for a planet to orbit the sun and the length of the axis of its ellipses. He had the vision that there must be a mathematical link between the planets' distances from the Sun and their velocities. But what was that link? Another IQ test was to be solved. What was the relationship between the sequence of numbers 58, 108, 150, 228, 778, 1430 (semi-axes of the planets' ellipses, measured in millions of kilometers) and the sequence 88, 225, 365, 687, 4392, 10753 (orbital periods, measured in days)? Kepler solved the problem with supreme ease. He determined that the square of the orbital period, divided by the third power of the semi-axis, is nearly exactly equal to 0.04 for all six planets. This time his intuition, again without the faintest trace of a justification, led to one of the most fundamental laws of nature.

Both Kepler's correct and incorrect insights originated from his deeply seated conviction that God had created the world based on numerical laws. The idea of nesting platonic solids, each encased in a sphere, within one another, seemed just as plausible to natural scientists during the Age of Enlightenment as that the square of one variable should be proportional to the cube of another. While the first hypothesis proved to be a figment of Kepler's fertile imagination, the second would go down in the annals as a path breaking discovery.

For many years, Kepler's three laws remained nothing but a numerical curiosity. The natural philosophers of the time, all of them men of faith, were convinced that the reason for the laws, if there was one, would surely remain a mystery within God's eternal wisdom forever. It was only in 1687 with Isaac Newton's monumental *Philosophiæ Naturalis Principia Mathematica* (The Mathematical Principles of Natural Philosophy) that Kepler's laws were put on firm theoretical footing. The English physicist provided mathematical proof that the motion of planets does not just obey divine rule but that the ellipses are an absolute necessity.

Newton's contemporaries did not take readily to his law of gravitation. While everyone understands that a cart moves once it is pulled by the towing bar, it requires much imagination to accept that the cart can be moved from a distance without even so much as touching the bar. Nevertheless, Newton's model still required a divine force which would act as a regulator, dealing with issues of stability and lost energy. It was only Newton's successor in France, the pioneer of theoretical mechanics, Pierre-Simon de Laplace, who would do away with the hypothesis of God as a necessary requirement to bringing order to the world.

In spite of his seemingly supreme rationality, the deeply religious Newton never stopped dabbling in esoteric sciences, the occult, and numerology. If Kepler had a weakness for astrology, it was alchemy for which Newton developed a passion and to which he would remain attached for the rest of his life. His nightly search for the legendary "philosopher's stone" remained fruitless, of course. If anything, the mixing and pouring of toxic substances may have caused him chemical poisoning, possibly from mercury. But the search for a method to manufacture gold was then considered *de rigeur*, even among natural scientists. Newton even taught himself Hebrew in order to read the Five Books of Moses in the original language. He covered thousands of pages with abstruse numerological calculations in attempts to extract scientific information from the scriptures. After having spent hundreds of hours unravelling God's secret laws as they supposedly manifested themselves in the Holy Bible, he was led to the inescapable conclusion that the world would come to an end in 2060..., and if not then, then surely by 2370.

Across the Channel, meanwhile, in Hannover, there lived Gottfried Wilhelm Leibniz, Newton's intellectual equal and opponent in every respect. Ahead of his time, intellectually, by several generations, Leibniz is famous for, among many other things, having developed the concept of a calculator based on the binary number system.

Leibniz was also on par with his Englishman adversary when it came to mysticism. For Leibniz the digits zero and one of the binary system were more than just a calculating device. They represented nothing less than the entry ticket to the understanding of genesis. One portrayed God, zero depicted the Void. The digital number seven represented the seventh day of Creation, the Holy Sabbath, which, in binary notation, is written as 111. This, of course, is a symbol for the Trinity, and so on. *"Cum Deus calculat, fit mundus,"* (while God calculates, the World is created), he wrote. Leibniz was convinced that he had not actually invented the binary system but merely discovered it. Hugely impressed by the binary system, he thought that with its help he could convert the Chinese—who already possessed the binary symbols Yin and Yang—to Catholicism.

The Pythagorean world view once again came into its own in 1869, when Dmitri Ivanovich Mendeleev presented his suggestion of a periodic table of chemical elements, arranged in order of their atomic mass. With wise foresight, Mendeleev left some spaces in his table empty for elements that were as yet undiscovered although very little hinted, at the time, that further chemical elements existed. In the gap between the elements zinc with the atomic mass of 30, and arsenic with the atomic mass of 33—both known since antiquity—there just had to exist elements with the atomic masses of 31 and 32. Mendeleev remained firmly convinced that these empty spots would be filled at some time. Only a few years later, he was proved correct when gallium and germanium were discovered and their masses matched his predictions.

Around the same time, in 1885, a Swiss school teacher by the name Johann Jakob Balmer was fascinated by the Kabbalah. Just by studying numerology he discovered a simple formula for the wavelengths of the spectral lines of hydrogen. It was left to Niels Bohr, thirty years later, to properly explain the causes for this phenomenon by means of quantum mechanics.

Carl Friedrich Gauss, the leading mathematical light of the late eighteenth and early nineteenth century, the Prince of Mathematics as he was later called, had been fascinated by numbers since his early childhood. Many anecdotes are known about the mathematical abilities of the young Gauss. He was able to do calculations well before he could even talk. As a three-year old, he corrected an error in his father's wage calculations and, at the age of eight he astonished his teacher by instantly solving a busy-work problem: to find the sum of the first 100 integers. As an adult he, of course, did much more serious work. With his masterful book *Disquisitiones Arithmeticae*, published in 1798, he single-handedly brought the study of number theory, then called higher arithmetic, to new heights. His famous prime number theorem, which would remain unpublished for many years, describes how prime numbers are distributed among the integers. Although Gauss remained a devout Christian throughout his life, his study of numbers had nothing in common with mysticism. For him, both God and number theory were complete and perfect, a belief which he summed up in his declaration "God does arithmetic".

Toward the end of the nineteenth century, Georg Cantor revolutionized the mathematical world. He created set theory which postulated different degrees of infinity. Jesuits used his concepts to derive the existence of a God to whom an exclusive claim to supreme infinity may be attributed. Cantor immediately distanced himself from such interpretation, of course. On the other hand, he got ensnared in daring theological speculations when speculating about the set of all sets—a concept for which even logic breaks down. It was little surprise then that his work did not receive universal acclaim and that opponents sought to ridicule set theory. Leopold Kronecker in Berlin summarized it thus: "God created the integers; everything else is the work of man". An American mathematician added that set theory, being a theory for God, is best left to God. At the other end of the range of opinions was David Hilbert from Göttingen, the most influential mathematician of the early twentieth century. He vigorously supporter Cantor, famously exclaiming that "Nobody shall expel us from the paradise that Cantor has created for us."

The many years spent studying objects that nobody had ever seen before him, and the hostilities he had to suffer, had an unfortunate effect on Cantor. Throughout his life, he was plagued by bouts of depression. He spent the last years of his life in a psychiatric clinic where he died in 1918. The controversy surrounding his set theory has never subsided.

Number mysticism was not universally derided by natural scientists, even until quite recently. Sir Arthur Eddington, one of the most eminent astrophysicists of the twentieth century was firmly convinced that the numerical values of the radius of the universe, its mass and its age as well as the speed of the light and the gravitational constant had to be in some harmonious relationship with each other, even though there was nothing whatsoever to justify his assumption.

Most of his colleagues smiled at the abstruse numbers game played by the aging scientist.

One who did not smile was Eddington's compatriot and fellow scientist, the physicist and Nobel Prize winner Paul Dirac. Quite enamored by the beauty of mathematics, he justified the Pythagorean world view with the claim that if there is a choice between two theories—one ugly, the other beautiful—the beautiful one would always win out, even if the ugly one fits the experimental data better. While Dirac thus threw the scientific method overboard, his quest to express nature in beautiful equations would nevertheless prove very useful. Seeking esthetically pleasing formulations, he came across an equation that neatly combined the theory of relativity with quantum mechanics. Unfortunately, the equation had two separate solutions, one of which seemed to be quite meaningless at first glance. But it was too beautiful for Dirac to just let go. His persistence paid off. The nonsensical, but beautiful solution was the first indication of the existence of anti-matter. "God used beautiful mathematics in creating the world," he stated with reverence. Meanwhile, Einstein thought he could detect human traits in God's universe which led him to conclude that "God is subtle, but malicious he is not."

Beauty in mathematics was also of high importance to the philosopher and Nobel laureate in literature Bertrand Russell. "Mathematics, rightly viewed, possesses not only truth, but supreme beauty— a beauty cold and austere, like that of sculpture," he wrote. G. H. Hardy, the Cambridge number theorist, resolutely claimed that "Beauty is the first test; there is no permanent place in the world for ugly mathematics." With this as standard, the most sublime expression of esthetics in mathematics must surely lie in the extraordinary formula which the Swiss mathematician Leonhard Euler proved in 1748: $-e^{i\pi} + 1 = 0$. In a sequence of symbols it links five fundamental mathematical constants through addition, subtraction, multiplication, and potentiation—namely, e, the base of the natural logarithms, i, the square root of -1, π, the ratio of the circumference of a circle to its diameter, 1, and 0—into one slim formula.

Number mystics want to comprehend the cosmos and predict the future on the basis of numbers. In this, they are not so very dissimilar to their brethren, the scientists. For many a discovery, it was Pythagorean intuition, rather than rational analysis, that was crucial—even if the brethren do not readily admit it. Nowadays, researchers possess the technical tools of statistics, such as regression analysis, to search in an objective manner for possible interrelationships. But these tools can be mis-used. So-called data miners, modern-day number mystics, correlate everything with everything in a search for possible relationships that, only in retrospect, may seem faintly plausible. The proper theoretical framework which would justify the claims is then supplied later, merely as an afterthought.

Meanwhile, the Pythagorean concept of the cosmos has reached a new level. Alan Turing and John von Neumann heralded the age of the electronic computer in the middle of the twentieth century. A few years ago, Stephen Wolfram presented *A New Kind of Science*, a book in which he claims that the entire universe is one giant computer producing complexity through the repeated execution of simple rules. Thus, mankind has travelled the voyage from the Pythagorean dictum "all is number" to its modern-day equivalent "all is computation."

Math for math's sake

Chapter 2

Getting Closer to the Roots

Among the mathematical constants, the number Pi or the Euler number e are known to be absolute stars. Many more constants exist, however, but they are less well known. One example is the Littlewood-Salem-Izumi constant, which plays a significant role in the geometry of right-angled triangles.

The so-called trigonometric functions such as sine, cosine, and tangent have numerous applications in physics, engineering sciences, and in surveying. Pure mathematicians, on the other hand, are more interested in the theoretical aspects of the functions; for example, what happens when trigonometric functions are multiplied by factors and then added. Does the sum approach a limit when successively more elements are added, or does it escape towards infinity? In the classic textbook *Trigonometric Series*, published in 1935, the Polish mathematician Antoni Zygmund demonstrated for a particular sum of cosines that the value of a certain parameter determines whether the sum is finite or not. If the parameter is larger than a certain value, the sum is finite; if it is smaller, the sum tends towards infinity. In his proof, Zygmund referred to previous, unpublished work by John E. Littlewood, Raphaël Salem, and Shin-ichi Izumi which is why this value is called the Littlewood-Salem-Izumi constant. In order to find the exact value of this constant, a certain integral must be set equal to zero and that is where the going gets tricky. The integral is not exactly solvable and can be computed only by approximation.

In 1964, when computer science was still in its infancy, two mathematicians from Northwestern University published a short article in the journal, *Mathematics of Computation*, informing the public that the value of the constant lies between 0.30483 and 0.30484. They had computed the 'root' of the integral, the place where it hits zero. They used approximation techniques on the 709 Data Processing System manufactured by IBM, one of the fastest and most advanced computer systems at that time. But just six months later, Robert Church from the Sperry Rand Center in Massachussetts poured cold water on their work. He submitted a short piece in which he alerted his colleagues that the result was faulty from the third decimal onwards.

Church's article was submitted on October 28, 1964 to *Mathematics of Computation*. Barely six weeks later, the mathematicians Yudell Luke, Wyman Fair, Geraldine Coombs, and Rosemary Moran from the Midwest Research Institute in Missouri not only confirmed Church's result but improved on it. By performing calculations on an IBM 162 scientific computer, they calculated the Littlewood-Salem-Izumi constant to 15 decimals.

A period of silence followed which lasted for forty-five years. In 2009 a paper was again published about the constant in *Mathematics of Computation*—albeit, keeping up with the times, in electronic form. In it, the Spanish mathematician Juan Arias de Reyna from the Universidad de Sevilla and his Dutch colleague Jan van de Lune from the Center for Mathematics and Computer Sciences in Amsterdam presented work in which they computed the constant once again, but to a much higher precision. They used a method that had been developed by Isaac Newton and Joseph Raphson in the seventeenth century to approximate roots of functions and integrals. Thus, with the help of an ancient method, implemented on the most modern of machines, they were able to calculate the constant on a laptop computer to a precision of 5000 decimals within 20 minutes.

One might well ask why mathematicians want to spend—some would say waste—their time with such calculations. Who benefits from knowing 5000 positions after the decimal point? Well, it was not just the more precise value of the Littlewood-Salem-Izumi constant that had aroused the interest of de Reyna and van de Lune. Rather, they explained, they were interested in the methods that could be utilized to perform the calculations. They wanted to demonstrate that all it takes to master some of today's mathematical challenges is a laptop, a modern commercial software programme such as Mathematica, and an approximation method dating from the seventeenth century.

Chapter 3

Mock Functions

The Indian mathematician Srinivasa Ramanujan (1887–1920) was not your regular, run-of-the-mill genius. Having studied mathematics only for a single year, he was self-taught, but—under the gentle guidance of his friend, mentor, and collaborator in Cambridge, G. H. Hardy—produced truly pioneering results during his unfortunately short life. Some of those results have kept several generations of mathematicians busy.

Two months before his death at age 32, probably due to tuberculosis, Ramanujan wrote a last letter to Hardy. "I discovered very interesting functions recently which I call 'mock' theta functions," he wrote. "Unlike the 'false' theta functions... they enter into mathematics as beautifully as ordinary theta-functions. I am sending you with this letter some examples... ." He gave his discovery the strange name because he believed that his newly discovered functions shared a certain similarity with the theta functions that had been introduced by Carl Gustav Jacobi at the beginning of the nineteenth century.

In his letter, Ramanujan presented seventeen mysterious power series. But he provided neither a definition, nor a way of constructing them, gave no hint as to why he considered them so significant, and did not even say what they had in common. The possibility remains that he revealed more precise information to Hardy but we will never know since the first few pages of his letter remain lost. However, since

Ramanujan possessed an uncanny sense for deep mathematical relationships, most mathematicians were convinced that some important theory must be hidden behind these functions.

After Ramanujan's death, his widow bequeathed her husband's notebooks, into which he had densely packed all of his 3,542 theorems, to the University of Madras. The university, in turn, forwarded the notebooks to Cambridge University and mathematicians have been scrutinizing them with fine toothed combs ever since, in the faint hope of finding further pearls. In 1976, one of them struck a bonanza. Working in the library of Trinity College in Cambridge, the American mathematician, George Andrews, chanced upon a bundle of notes in Ramanujan's handwriting, 138 pages in all, that nobody had ever seen before. They contained the Indian mathematician's output of the last year of his life. The find soon became known as *Ramanujan's lost notebook*. According to one expert, the discovery of this notebook "caused roughly as much stir in the mathematical world as the discovery of Beethoven's tenth symphony would cause in the musical world." When examining this notebook, mathematicians found two further mock theta functions. (In the 1930s a British mathematician had independently discovered an additional three such functions.)

In subsequent decades, the mysterious power series became useful in such diverse areas as number theory, probability, combinatorics, mathematical physics, chemistry, and even cancer research. But no progress was achieved in understanding them. Mathematicians proved theorems about the use of mock theta functions, without any real idea about what these mysterious objects really were. But the many applications both within and outside of pure mathematics made it obvious that the functions just had to be part of an important theory. It only waited to be discovered.

The first breakthrough occurred in 2002 when the Dutch mathematician Sander Zwegers proved that a mock theta function is one part of the so-called real-analytic modular forms. The latter play an important role in number theory (for example in the proof of Fermat's Theorem), algebraic topology, function theory, or string theory. But the essential question, how these functions derive from an overarching theory, still remained unanswered.

This is where Kathrin Bringmann and Ken Ono of the University of Wisconsin-Madison make their entry. In a series of path-breaking papers, the two mathematicians showed that mock theta functions belong to a new theory which links classical modular forms with so-called harmonic Maass forms—a modern generalization of modular forms. With this, Ramanujan's condundrum was at long last solved. Additionally, the papers showed that there exist not only the fewer than two dozen mock theta functions that were known, but an infinite number of them. The importance of these papers can be gauged by the fact that with the help of their theory, Bringmann and Ono were able to prove some longstanding conjectures in number theory.

Chapter 4

Meanderings of a Mathematical Proof

The proof of Kepler's Conjecture, which was four centuries in the making, continued to preoccupy the minds of many mathematicians even six years after it had been presented to the world for the first time. In 1611 Kepler had suggested that the most efficient way to pack spheres is to stack them in a pyramidal fashion, as the owners of grocery stalls do all over the world with tomatoes, apples, and oranges. But proving this thesis which seems so obvious turned out to be difficult beyond expectation. Only in 1998 did Thomas C. Hales, then professor of mathematics at the University of Michigan, succeed. He proved Kepler's Conjecture with the help of computers. Such an approach to proving theorems, rejected by many in the mathematical community, is dubbed a brute force approach.

Robert MacPherson from the Institute for Advanced Study at Princeton University, editor of the renowned journal, *Annals of Mathematics* wanted to publish the proof. As is customary with all scientific articles before they are published, he initiated a rigorous review of the paper. A dozen mathematicians pored over hundreds of pages and reams of computer output, scrutinizing and questioning every little detail. But after five years, they threw in the towel. While they found no errors, flaws, or bugs, they still felt uncomfortable. It was just impossible to check every line of computer code and retrace every step of the computers' computations. Thus the proof's correctness could not be verified with absolute certainty. Exasperated and tired, they wrote that they could not guarantee that the computer-assisted proof was absolutely correct.

Instead of rejecting the paper that had, in the meantime, become very famous, the Annals' editors wanted to take a middle road. They decided to publish the proof anyway, but to add a disclaimer alerting readers to the inherent problems of computer-assisted proofs. This did not go over well with many colleagues. A mathematical proof is either correct or it is not. There just are no gray areas in mathematics. Adding a disclaimer would do little else than cast doubt on the work of Tom Hales and his student, Sam Ferguson.

Such a predicament calls for a Solomonic decision. The editors arrived at it by splitting the paper in two. One part, published in the *Annals of Mathematics* in November 2005, only describes the strategy of the proof. A second version of the proof, which also includes the more contested bits, was published as a series of six articles in the journal *Discrete and Computational Geometry* in 2006.

While the Annals of Mathematics may claim to have found a way out of a predicament, its approach was not entirely consistent. In 2003, after a seven-year waiting period, the Annals published an article by three mathematicians which describes how computers went through a billion computations in order to identify seven special cases of a problem. Doron Zeilberger from Rutgers University, an advocate of computer proofs, thinks that the journal sometimes employs double standards. He concedes, however, that when dealing with the proof of something as old and prestigious as Kepler's Conjecture, the bar must be set somewhat higher.

Chapter 5

All Roads Lead to Rome

Nowadays, it is common practice that a driver, lost in a maze of streets, gets instructions from his friend over a mobile phone on how to get to his destination. All he needs to do is announce his current position and the friend can talk him through the ride. Wouldn't it be nice if the friend were able to help the driver, even if the latter has no idea where he is. Let us say the driver lost his way and cannot read the street signs because they are in a foreign script. Let us assume further that all streets are marked with two colors. The instructions could then be: "take the green street until the next intersection, then take a green street again and then the blue one. Once you get to the next intersection, take the green one again, again the green one, and then again the blue street. Just keep going like that until you arrive at my place." The chief attraction of the matter is that by following the green- green- blue instructions over and over again the driver will inevitably get to his destination, no matter where he started out. The friend does not even need to ask him where he is, before giving the instructions.

Are there mazes for which such instructions always work? The problem, commonly referred to as the "street-coloring problem," is a question from the theory of graphs. A graph is a net consisting of a collection of nodes, and a collection of edges that connect these nodes. Graphs are used to model real-world systems such as the Internet (nodes represent routers, edges represent connections between them), airline networks (each node is an airport and each edge is the leg of a flight), or city road networks (nodes represent intersections and the edges are the streets leading from one to the other).

We consider a network in which each node has the same number of outgoing edges leading to other nodes. Furthermore, each node must be reachable from each other node, possibly by a route meandering via many different nodes. The question is whether the edges of such a network can be colored in a manner such that a simple, repetitive set of directions always leads to the destination. In 1970, two mathematicians conjectured that all networks that fulfill a further, rather technical, condition can, indeed, be colored in this way. (The technical condition is that the number of edges of all cycles through the graph—loops that lead back to the same node—must be "mutually prime." This means that if there is a cycle consisting of three edges, the other cycles in the graph must not consist of 6, 9, 12 etc. edges.)

The two researchers could not prove their conjecture, however, and the problem fell into near oblivion for about forty years. Every once in a while, some mathematicians found partial answers for specific networks; since 1970 a total of 16 publications about the problems appeared. But up until recently, it was not known whether the conjecture was generally correct.

Then, in 2008, news came quite unexpectedly over the Internet that the conjecture had been confirmed. The Israeli mathematician Avraham Trahtman had proved that all graphs fulfilling the technical condition could be colored as described. The result is not only an intellectual achievement in pure mathematics but may prove to be of practical significance for computer science. It shows that errors that occur when entering data or other disturbances may not be important under certain circumstances. Just as the driver who got lost in the maze of streets does not need to know where exactly he was when he placed the call for help, a computer program could be led back from an incorrect state to the correct state with the help of simple, repetitive instructions.

Of course, the proof that a coloring exists enabling a driver to find the right path through a net, is only the first step. The next step would be to find out which edges need to be colored in which color. Recently, two French mathematicians presented a computer algorithm which allegedly is capable of calculating the appropriate coloring of networks within a reasonable amount of time.

The personal history of Avraham Trahtman is extraordinary. The Russian mathematician, today in his mid-sixties, originates from Sverdlovsk (today Ekaterinburg) in the Ural Mountains. He studied mathematics at Ural State University and then taught at the Technical University of Sverdlovsk for fifteen years. He lost his position in 1984 for "anti-Soviet activities". His crime was that he wrote an open letter of protest to the then Secretary of the Communist Party of Sverdlovsk, a man by the name of Boris Yeltsin.

For seven years, Trahtman could keep himself barely above water by working odd jobs as a programmer. It was only much later that he was able to assume, once again, a teaching position at a pedagogical institute. In 1992, he emigrated to Israel. With the fall of the communist regime, a million Jews left the Soviet Union and came to Israel. Among them were numerous extremely talented mathematicians who found it difficult to find work at one of the country's seven universities. Unfortunately, Trahtman who had a very respectable list of mathematical publications, was among those who could not find employment in their field. For two years, he worked as janitor, substitute teacher, and watchman. Finally, in 1995, Trahtman was at long last offered a teaching position at Bar Ilan University in Tel Aviv.

Chapter 6

Secrets Hidden in Numbers

Any primary school child is able to multiply or divide fractions. Therefore it is not unreasonable to assume that everybody knows whatever there is to know about these basic arithmetic operations. Nevertheless, higher mathematics still bothers to deal with this subject. Take as an example the sequence of fractions 1/2, 2/3, 3/4, 4/5 and so on until $n/(n+1)$. Now take a random selection of these fractions, multiply them or divide them, just as you wish. The result is, again, a fraction which is then simplified as far as possible. The question now is, how big can the largest numerator be that is obtained in this manner?

Several mathematicians have studied this and similar problems in the past. In 2005, the French mathematicians Régis de la Bretèche and Gérald Tenenbaum, together with their American colleague Carl Pomerance, presented yet another study. It was published in the April edition of that year's *Ramanujan Journal*.

The said journal is named after Srinivasa Ramanujan, a simple Indian accounting clerk turned brilliant mathematician. Born in 1887 into a devout Brahmin family in a small village, 400 kilometres from Madras, he already showed unusual mathematical skills early on. Without any higher formal education, he acquired the tools of the trade all on his own by just reading and mastering mathematical books which he borrowed from others. Many years he toiled away in India at jobs which did not even come close to meeting his potential. Often sick, always poor, he devoted all his spare time to the study of mathematics. Finally, Ramanujan gathered his courage and drafted letters to leading British mathematicians, attaching some of his results. But his methods were so novel and his presentation so unclear that most recipients, all of them well-known mathematicians, simply dismissed the letters. They thought that this Indian clerk was either a fraud or a crank. The writer simply lacked the educational foundations to be taken seriously. But one professor did take notice. It was Godfrey Harold Hardy of Cambridge, a leading mathematician at the time. He immediately saw that Ramanujan's theorems were highly intriguing and quickly realized that this young and unknown clerk from India was an outstandingly gifted mathematician, "a man of altogether exceptional originality and power."

Hardy immediately extended an invitation for Ramanujan to come to Trinity College in Cambridge. But it took several years to convince the strictly orthodox Indian to travel to England. As a Brahmin, overseas journeys were forbidden to him. Eventually, the prospect of studying under and with the world's best mathematicians got the better of him and in 1914 Ramanujan booked a passage to Cambridge. A most fruitful collaboration developed between him and Hardy, with the latter carefully steering his protégé toward mathematical rigor without stifling his intuitive approach to mathematics. The partnership between the two highly contrasting personalities lasted throughout the six years that Ramanujan spent in England. The two men could hardly have been more different in their beliefs, working styles, and cultures. Yet Hardy, an atheist with a traditional belief in rigorous proof, and Ramanujan, a deeply religious man who relied heavily on his intuition, developed a deep relationship. For Hardy it was "the one romantic incident of my life". The Cambridge Don was able to fill the gaps in Ramanujan's education, always mindful not to disrupt the young Indian's spell of inspiration. Ramanujan did not disappoint. In barely six years, he achieved astounding results under Hardy's tutelage. His ability to discern secrets that were hidden away in certain numbers or series of numbers was legendary.

Ramanujan became one of the youngest members in the history of the Royal Society and the first Indian to be elected a Fellow of Trinity College.

But the Indian remained lonely in Cambridge, suffering from stress and homesickness and maybe also from the lack of vegetarian food which was the only form of nourishment his religious beliefs allowed him to consume. After all, fresh produce was scarce during the first World War. In 1919 he decided that he could take it no longer. Already weakened by sickness, he returned to India, where his wife had remained throughout his absence. Soon thereafter he died at the age of just 33. His legacy is enormous. Even today researchers are still ploughing through his notebooks desperately hoping to come across more hidden treasures.

The Ramanujan Journal, founded in 1997, appears four times a year and is devoted to those areas in mathematics which were so profoundly enriched by this extremely talented Indian. It is undoubtedly the appropriate outlet for articles on problems of higher arithmetic, such as the one mentioned above. At the outset of their paper, the authors remark that in any case, the largest numerator must be smaller than $1 \times 2 \times 3 \times \ldots \times (n+1)$. This is, however, a very broad upper bound. For $n = 15$ it already reaches a value of 20 trillion. All we know is that the numerator must be smaller than this upper bound. This knowledge is not very useful and reinforced the necessity to restrict the bound on the numerator's maximal value. The three mathematicians proved that for $n = 20$, the maximum value of the numerator lies somewhere between a trillion and 100 quadrillion—a rather wide range. For the set of the first thousand fractions ($n = 1000$) the numerator's maximum value lies between the unimaginably large number 10^{600} and 10^{2000}. Ambitious number-theorists, eager to carve out a niche for themselves, may try to establish narrower bounds for such intervals.

Before we end, let me make a brief comment to dispel the erroneous assumption that mathematicians are narrow-minded scientists: Gérald Tenenbaum, one of the paper's three authors, spends his time not only teaching and doing research at the Université Henri Poincaré in Nancy, France, but has also made a name for himself as an author.

He published a play for the theatre "Trois pieces faciles" (Three Easy Pieces) and two novels, "Rendez-vous au bord d'une ombre" (Date at the Edge of a Shadow) and "L'ordre des jour" ("Order of the Days"). These three pieces of literature should put to rest the stereotype that with mathematicians you can talk about nothing but numbers.

Chapter 7

Prime Time for Primes

In 2004, two mathematicians published a paper on the Internet, in which they confirmed a hitherto unproven conjecture regarding prime numbers. They proved that arbitrarily long arithmetic sequences exist, all of whose elements are primes. An arithmetic sequence is a progression of numbers of the form $a + bk$, where a and b are fixed integer numbers and k takes on integer values between zero and an arbitrarily high upper bound. If all elements of this sequence are prime, we speak of an arithmetic prime number sequence. The sequence 5, 11, 17, 23, 29, which can be written as $5 + 6k$, with k running from zero to 4, is an example of an arithmetic prime number sequence. The longest arithmetic prime number sequences known to us today consist of 22 elements. One of them is of the form 11,410,337,850,553 + 4,609,098,694,200k.

As far back as 1770, arithmetic sequences of primes were being studied by Joseph-Louis Lagrange in France and Edward Waring in Britain. Two questions were of interest. Are there infinitely many arithmetic prime number sequences of a certain length? And, do sequences exist which are arbitrarily long? In 1939 the Dutch mathematician Johannes van der Corput proved that infinitely many arithmetic prime number sequences of length 3 exist. All else remained unknown. While there was a strong suspicion in the mathematical community that arbitrarily long arithmetic prime number sequences do indeed exist, a proof to that effect was missing.

Then Ben Green, a 27-year old Cambridge mathematics graduate, and his 29-year old colleague Terence Tao from UCLA—also known as the mathematical Mozart after he gained his Ph.D. at the age of 21—stepped onto the scene. They answered both questions in the affirmative: for any given length, an infinite number of arithmetic prime number sequences exist that are longer than that length.

Green and Tao decided to first tackle the question of arithmetic sequences consisting of four prime numbers. Their approach was to embed primes in the set of "almost primes". The latter are numbers which are the product of a few prime numbers. This made their work much easier since suitable mathematical tools existed to work with almost primes. But they soon got stuck. As Tao recounted, they were forced to estimate the size of some "unsavoury" expressions. In the attempt to master this difficulty, Tao and Green realized that their procedure thus far bore some similarity to the so-called ergodic theory—a sort of statistics, inspired by physics. This insight induced them to change direction, which permitted a simpler treatment of arithmetic sequences consisting of four primes. More importantly, the new thrust allowed Tao and Green to extend their proof to arithmetic sequences of primes of arbitrary length.

But research is rarely smooth sailing and before Green and Tao could reach their breakthrough they found themselves in rough waters. They were faced with remainder terms which had to shrink to zero for sequences of arbitrary length. The hurdle seemed insurmountable. It was a chance encounter with Andrew Granville, a British mathematician working in Canada, that came to their rescue. Granville told them about an error that he had discovered a year previously in the alleged proof of the so-called Twin Prime Conjecture by Daniel Goldston and Cem Yıldırım. What had happened was that Goldston and Yıldırım had glossed over...a remainder term. But one man's misfortune is another man's gain. Goldston and Yıldırım's failed approach provided Green and Tao with the necessary groundwork for an attack on *their* remainder term. It proved successful. They could now make the correct estimation. (Goldston could at least take some comfort in that his initial comment upon completing the purported proof—"I think some interesting math is going to come out of this, whatever the outcome"—turned out to be correct.)

A word of warning is in order: the term "arbitrarily long" must not be confused with the term "infinitely long". The former only signifies that for any given upper bound, arithmetic sequences of primes exist that are longer than this bound. The reason that no *infinitely* long arithmetic sequences of primes exist, can be seen from the following: the arithmetic sequence $a + bk$ (a and b constant, $k = 0, 1, 2, \ldots$) will, at the very latest, contain a composite number when arriving at the element $k = a$. Then we have the element $a + ba = a(1 + b)$. This number is divisible by a and by $(1 + b)$. Hence, it is not prime.

Tao and Green's fifty-page, very technical paper does not suggest, however, that arithmetic sequences of primes with more than 22 elements are about to be discovered. Their proof is nonconstructive, which means that it only proves the existence of arbitrarily long sequences, not how to find them.

Math applied to real life

Chapter 8

Stamps and Coins

The feeling was familiar, at least to those of us who were around before email became available. You had just written a letter, tucked it into an envelope and sealed it, when you remembered that the post office has already shut for the day. Luckily there was a supply of stamps in your drawer, ready for exactly this type of situation. With this, the immediate problem had been resolved. But where a problem ends for mere mortals, it just begins for mathematicians. Given the denomination of the available stamps, they asked themselves, what is the highest postage that fits onto an envelope of a given size? For example, with stamps of denominations 1, 4, 7, and 8 cents, all postages between one and 24 cents can be affixed to an envelope which has room for no more than three stamps.

The so-called 'stamp problem' dates back to 1937 when the German number theorist Hans Rohrbach (1903–1993) first described it in an essay. Ever since, many studies have addressed it and even today, articles are published which illustrate different aspects of the problem. It turns out that the problem of putting together a certain postage with the least number of stamps is not easy. In fact, the problem is very complex, as has been shown by Jeffrey O. Shallit from Waterloo University in Canada. He demonstrated that the time a computer would require to compute the optimal configuration of stamps would grow beyond all bounds with increasing postage.

In an article published in the *Journal of Integer Sequences*, the Indian mathematician Amitabha Tripathi investigated a special case of the stamp problem. He postulated that the denominations of the stamps increase by a fixed amount. With an increase of, for example, 7 cents, there would be stamps of denominations 1, 8, 15, 22 cents, etc. Tripathi developed a formula which gives the highest amount up to which all postages can be paid solely with a certain number of these stamps. Hence, with at most ten stamps of the above four denominations, all postages up to 94 cents can be put on the envelope.

The version of the problem without any restriction on the number of stamps is called the 'coin problem'. It is associated with the German mathematician Ferdinand Fröbenius (1849–1917) who introduced it in the context of paying for a purchase with exact change, given the availability of coins of specific denominations. In contrast to the stamp problem, it is the lower bound that is of interest here. Starting at which amount can any purchase be paid for, with the available denominations? In a letter to the editor of the Educational Times in England, the British number theorist James Joseph Sylvester (1814–1897) gave a solution to this problem. If there are just two coins, A and B, which share no common positive divisor except 1 (hence, they are 'relatively prime') all prices higher than $A \times B - A - B$ can be paid. If, for example, only five-cent and two-cent coins are available, everything that costs four cents or more can be paid. With five-cent and 13-cent coins, one could only pay for all purchases of 48 cents or more (seven 5-cent coins and one 13-cent coin). Below 48 cents there are many gaps. There exists a computer program that finds the lower bounds for three coins of different denominations. For four coins or more, the answers are unknown. There exist only estimates.

Still a different version of the stamp and the coin problems is the "McNugget problem". The reference is to McDonald's Chicken nuggets which are sold in boxes of six, nine, and twenty nuggets. So-called 'McNugget numbers' are the amounts of nuggets that can be bought (and consumed) by combining any number of boxes. For example, 44 nuggets can be bought by purchasing one twenty-nugget box, two nine-nugget boxes and one six-nugget box. But no combination of boxes will result in a total of, say, 13 or 22 or 37 nuggets.

The question is, which is the largest non McNugget number, i.e., the largest number of nuggets that cannot be purchased as a combination of boxes. It turns out that it is 43. Any number of nuggets beyond that can be purchased. When the Happy Meal box, containing four nuggets, was added to McDonald's menu, the largest non McNugget number dropped to 11.

Problems related to the stamp problem, the coin problem, and the McNugget problem are the most efficient combination of coins when giving change and the optimal denominations for a country's currency. In the U.S. there are five different coins: pennies (1 cent), nickels (5 cents), dimes (10 cents), quarters (25 cents) as well as the rarely used half-dollar (50 cents). Under the assumption that all prices are equally probable, store owners today require 4.7 coins on average in order to give change. Jeffrey Shallit calculated that the required number of coins for change would decrease by 17 percent if, instead of the dime, a coin of value 18 cents were minted. In Europe too, the addition of a 1.33 Euro coin would reduce the required number of coins for change from an average of 4.6 to 3.9. We would rather not think of the confusion at the checkout counters.

Chapter 9

On the (Un)Fairness of Queues

Nobody likes to queue—this includes mathematicians even though they could, if they so chose, put their insider knowledge about queues and queuing to practical use. Queuing is a waste of time, especially during skiing vacations. A case in point is the introduction of a modern six-seater express chairlift in the Swiss alpine resort of Flims. Several years ago, it replaced the old twin-seater chairlift and the T-bar, both built in the sixties.

The old lifts had a combined capacity of 3450 passengers per hour, while the new, super-modern construction alone transports 3200 passengers per hour...provided it functions without any hiccups. But in its first year of operation disruptions invariably occurred and the smooth running of the lift with the expected maximum load of passengers failed to materialize. When the lift first began to operate, a maximum of only 2700 passengers could be transported up the mountain per hour which hugely disappointed the construction firm as well as the resort's tourist board and, of course, the skiers.

For the latter, the bumpy start meant that more skiers arrived at the bottom of the lifts than could be taken up. As a consequence, a queue started to form at the bottom station which grew by the minute. Toward the late morning, skiers already had to endure a 30-minute wait. If many of them had not decided at this point to skip that particular mountain for the rest of the day, the queue would have grown by about 500 people per hour.

Of course there is a mathematical theory for these tedious queues. It was first developed by Agner Krarup Erlang, a Danish mathematician who worked in the early 1900s at the Kopenhagen Telephone Company. He was charged with the problem of determining how many circuits and how many operators would be required to provide acceptable telephone service. Upon examination of the number of telephone calls which might be made at the same time to the switch board stations, he developed a formula which computed the probability that all circuits are busy simultaneously. Refined by subsequent developments, Erlang's equation, which also allowed the calculation of the average waiting time, is still in use today, for example when call centers estimate the number of required phone lines.

For the uninitiated, queues are an unpleasant nuisance, no matter what. There is no good way to describe them and no good way to get around them. Or is there? Is every queue the same? It is not, at least to the expert who recognizes subtle and important differences which distinguish one queue from another. One feature of a queue is the time distribution of customers joining the queue. Are arrival times random and independent of each other, similar to cars driving up to a toll booth? Or do customers appear in specific rhythms, similar to travellers at airport passport control whose bunched arrival times depend on when an aircraft lands.

Hence, there is more to queues than a horde of miserable people waiting to be served. One important characteristic of queues is queue discipline. In countries where people abide by the traditional codes of conduct, queue discipline determines the order in which arrivals are serviced. For example, customers can be attended to in the order of their arrival, like car drivers setting off after a traffic light has changed from red to green. But customers can also be served in reverse order, similar to people getting out of an elevator. Still a different, but clever way of servicing waiting customers is to attend to those first whose business takes the least amount of time. This would minimize the total combined waiting times of all customers.

In 1961 John D. C. Little, a professor of marketing at MIT, developed a mathematical law which, although seemingly trivial, turned out to be so important that it was soon named after him. Little's Law states that the average number of customers in a queue is equal to their average arrival rate, multiplied by the average period of time that they remain in the system. If 60 people arrive per hour, and if ten minutes are required to deal with each of them, then on average there will always be ten people in the system. It is remarkable that this law holds, independent of the distribution of arrival times, service times, and queue discipline.

Since queues involve human beings and their respective behaviour, waiting time is not the only variable that needs to be considered. Psychological factors play an important part as well, though these are largely ignored by mathematicians. An airport is an obvious place for observations and Zurich with its two terminals is a case in point. At Terminal 2 several check-in counters are available and passengers naturally join the shortest queue. Bad luck it is, if a passenger in your queue happens to have a special request or does not have the right ticket. All people standing in this particular queue must wait while time-consuming arrangements are being completed. Meanwhile they can watch in frustration as the other queues move swiftly along.

Terminal 1, on the other side of the airport, has a different set up: one single queue feeds all the check-in counters. This results in shorter average waiting times than in Terminal 2. Surprisingly, it turns out that this would not necessarily entice passengers to chose Terminal 1 over Terminal 2, if they had a choice. The phenomenon is examined by four scientists from the Technion, the technological institute in Haifa, Israel. They found out that most passengers assume, incorrectly, that waiting times in a system which involves several queues is shorter. But the study also reveals a paradox. In spite of the perceived longer waiting times, the majority of the surveyed passengers prefers the single queue. The authors of the study believe that this is due to the fairness intrinsic in a single queue. Apparently, even-handedness is more important than waiting time.

Chapter 10

Run or Walk on the Walkway?

Terence Tao, a mathematician at the University of California in Los Angeles, is one of modern mathematics' greatest stars. As a teenager he was known as a prodigy, winning a gold medal at the Mathematical Olympiad at the age of twelve. (Previously, he had already won bronze and silver.) Promoted to full professor at UCLA by the age of 24, he won the Fields Medal at age 31.

Now, Tao has also become a blogger. He mostly holds forth on esoteric mathematical topics, using his blogs to expound on details from his lectures. But his blogs are not only meant for colleagues or advanced students, they also explore the mathematical background of daily occurrences and observations. This is the beauty of mathematics: it finds application in all sorts of daily events and common places.

On one occasion, Tao found himself late for his flight and had to rush from the check-in counter to the boarding gate. As in most airports, on part of the corridor between terminal and boarding area a moving walkway helps speed the passenger along; the remainder of the corridor has to be walked on firm ground. Not one to leave a good opportunity of exploring a mathematical challenge unexploited, Tao blogged the following question to his readers: if the passenger has to pause for a while in order to tie his shoelaces, should he do so on or off the moving walkway? Further, if he only has a limited amount of energy to do a short burst of running, say 20 seconds, and walks the rest of the way at a constant speed, is it more efficient to run while on the moving walkway or off the walkway?

When Tao blogs, the world blogs back. Within a few days, Tao was flooded with responses; comments and calculations abounded. Of course, there were also the usual jokers. Some suggested that it would be best to continue on the walkway with laces untied until you reach the belly of the airplane; others recommended avoiding running because one is likely to hit other passengers; still others cautioned that it would be dangerous to tie a shoe on the walkway since the laces could become entangled in the mechanism. One simpleton stressed that one always walks on the walkway, if for no other reason than its name. (This begs the question, however, what the aircraft is doing on the runway.)

But there were some serious comments as well. Some readers felt that it made no difference where the laces are tied and where one does the running, since there just had to be trade-offs in the times gained and lost. Others multiplied, divided, added, and subtracted the times spent walking, running, and tying shoelaces, and compared the results with the velocities of the walkway and the walking and running speeds.

Naturally the correct calculations gave the correct answer—which we will give away in a moment. But some nonmathematicians wished for a verbal explanation of the solution rather than an answer expressed in mathematical formulation.

One mathematician obliged. He asked the readers to imagine a pair of twins, Albert and Bertie, who reach the moving walkway simultaneously, each with one untied shoe. Just before stepping onto the moving walkway, Bertie bends down to tie his shoe. An instant later, just after getting on the moving walkway, Albert does the same. They both tie their laces, get up at the same time and continue to walk. Albert is ahead of Bertie, of course. After Albert steps off the walkway, Bertie, still on the walkway, does in fact close in on Albert. But after getting off the walkway he will never quite catch up, and it is Albert who made the correct choice. So here we have the correct answer to the question: tying the shoelace on the walkway is more efficient.

There is still the issue of where one should do the running. Tao's blog had also caught the attention of economists who felt that they should get involved. After all, optimizing variables—profits, market share, production—is their daily bread and butter. They concluded that in order to cover a distance in as little time as possible, one should spend as much time as possible on the fast section. If rushed passengers run on the moving walkway, they shorten the time spent on the fast section of the corridor and, at the same time, lengthen the time spent on the slow section. This is the reason why people should run on firm ground. In economic theory it has long been known that, given the choice, a production plant should schedule as little work as possible on inefficient machines.

The conclusion has implications for car races, believe it or not. Recently, Formula-1 drivers have been permitted to press a 'boost' button, releasing an 80hp burst of energy, for at most 6.6 seconds in every round. Tao's blog would advise racing car drivers to press the button during the slow stretches of the track.

Meanwhile, findings by Manoj Srinivasan, a locomotion researcher at Princeton University, showed that people slow down on walkways to conserve energy. The findings, published in *Chaos* in 2009, back up earlier results by Seth Young of Ohio State University who had observed that people walked slower on the automated walkways. The reason, says Young, is that when a person steps onto a moving walkway, the eye picks up that they are going faster than normal "leg tempo" and they naturally adjust to a more comfortable, viz., less energy-consuming speed.

Chapter 11

Suspicious Use of the Digit "9"

Common sense is a tricky thing—especially in mathematics. One would think that prices on the stock market should just as often begin with the digit 1, as with the digit 2 or, for that matter, with any other digit. Hence, surely every digit between 1 and 9 should occur as the leading digit in stock market prices in 11.1 percent. Similarly, we would assume that in collections of numbers which represent the populations of cities, physical or mathematical constants, the GNPs of all countries, no numeral should be preferred over another as a leading digit. But this assumption is untrue. In many sets of physical or social data the numbers are, in fact, distributed differently.

The first to take note of this surprising fact was Simon Newcomb, a Canadian astronomer and mathematician, who observed some 120 years ago, that the earlier pages of books of logarithms, used in pre-electronic calculator times to carry out arithmetic calculations, were far more worn-out than the later pages. From this he concluded that his colleagues must be dealing far more often with numbers starting with the digits 1 or 2 on the early pages, than those which start with an 8 or a 9 on the later pages. No sooner had Newcomb formulated the principle, than it was forgotten. It was not until 1938 that Frank Benford, an American physicist, re-discovered it.

But Benford took the observation a step further. He spent the following years collecting data to show that the pattern was very widespread. He tested sets of numbers from widely disparate categories: atomic weights, baseball statistics, areas of rivers, numbers appearing in magazines like the Reader's Digest. Every time he arrived at the same result: about 30 percent of the numbers start with a 1, about 18 with a 2, twelve with a 3, and so on. Less than 5 percent of the numbers possess a leading 9. The peculiar phenomenon was soon called the Benford distribution. Usually a quite accurate picture of Benford's distribution can be obtained by observing the day's stock quotations on the financial page of the local newspaper. The phenomenon is, by the way, quite independent of the currency. One gets similar results when translating the stock prices into Swiss francs or Japanese yen.

Benford's distribution is ubiquitous; it appears nearly everywhere one looks. In the American census of 1990, for example, it turned out that the populations of 3,000 districts were consistent with his law. But until an explanation for the phenomenon was discovered, the phenomenon remained no more than a curiosity. It was only in 1995 that the riddle was solved by Theodore Hill, a distinguished West Point graduate who went on to teach mathematics at the Georgia Institute of Technology. Hill placed the peculiar distribution of leading digits on firm ground. To explore his proof in detail would take us too far afield, so it should suffice to illustrate the phenomenon: let us think of a stock whose initial value is 100 dollars. The first digit is a 1. Assume now that its value increases throughout the year at a rate of 10 percent annually. It would take 88 months for the share price to reach the value 199. During that time the share would continue to have 1 as the leading digit. But it would then take only 52 months for the share price to rise from 200 to 299. During that time, stock quotations show a leading 2. And on it goes. The share would exhibit 9 as a leading digit only for 12 months since this is the time it takes for the share to rise from 900 to 999. Then the whole process starts over again: it will take 88 months for the share's value to rise from 1000 to 1999 and during that time the share, once again, has a leading 1.

This corresponds more or less to the distribution observed by Benford. Put mathematically, the leading digits of many real-life sets of data—where the data derives in some way or another from a growth process—are distributed logarithmically, rather than uniformly, as one might have expected. Then the frequency f of the leading digit d is given by the formula $f = \log(1 + 1/d)$. For $d = 1$, we get $f = 0.301$. For $d = 9$, the result is $f = 0.046$.

Once this phenomenon had risen from mere curiosity to the lofty heights of a mathematical law, the experts started to seek applications for it. An accounting professor, for example, scrutinized the leading digits of 170,000 tax declarations and examined whether the leading digits reflected the Benford distribution. In general they did, and when they did not it was usually due to erroneous or even fraudulent accounting data. No wonder then, that the IRS started to use Benford's law to identify tax fraud. For computer scientists too there was something in this. The fact that small digits occur more often as leading digits than large ones could possibly be used to design a more efficient computer architecture.

Chapter 12

The Letter Writers

Charles Darwin (1809–1882) and Albert Einstein (1879–1955) were prolific letter writers. 7,591 letters sent by Darwin and 6,530 letters which were received by him are known to us today. Einstein's correspondence contains at least 14,500 sent letters and 16,200 received letters. The availability of so much good data just begs for an analysis which the physicist Albert-László Barabási of the University of Notre Dame in Indiana and his student João Gama Oliveira happily supplied. Analyzing the time spans it took Darwin and Einstein to respond to specific letters, they found that the two intellectual giants' patterns of replies were similar to the reply-patterns of today's computer users who receive and send e-mails. But soon doubts were raised about the quality of the study.

Einstein answered about one quarter of the letters sent to him, about half of these within ten days. Darwin answered one third of his mail, two thirds within ten days. There was one notable outlier in the data: one correspondent had to wait 30 years for Darwin's reply. The more interesting point of the study, however, is the behaviour between these two extremes. Barabási and Oliveira found that the length of time it took the two scientists to reply can be best described by a so-called scaling law: the probability that an answer is sent within exactly τ days, $P(\tau)$, follows a power law, $P(\tau) \approx \tau^{-\alpha}$. Surprisingly, the value of the parameter α is very similar in both cases: $\alpha = 1.45$ for Darwin and 1.47 for Einstein.

The scientists were even more surprised when they became aware of the similarity to the pattern of today's e-mail correspondence. In earlier research, Barabási and colleagues had found that the e-mail correspondence of randomly selected people also follows a power law. According to the authors, this can be explained by the fact that people prioritize their mail, be it on paper or electronic, according to its importance. Then they answer it in accordance to "queuing theory", responding faster to high-priority mail than to low-priority mail. One consequence of queuing theory is that the distribution of response times follows the said scaling law.

Reactions towards the study, published in the highly respected journal *Nature*, ranged from reservation to outright rejection. Luis Amaral from Northwestern University in Evanston, Illinois, regards the conclusions as downright nonsense. With two of his colleagues, he analyzed Barabási's earlier study and concluded that a so-called log-normal distribution did a much better job at describing the observed distribution of waiting times. Significantly, such a distribution cannot arise due to queuing. Then the editor of *Nature* added oil to the fire by admitting that the empirical data used by Oliveira and Barabási had apparently been smoothed without this fact having been reported in the paper.

Another conclusion begs clarification. For e-mails, the parameter α which describes the waiting times between receipt of a message and its answer, is only 1.0. According to Oliveira this implies that, in principle, Einstein and Darwin replied to their correspondents faster than today's e-mail senders do. He explains this counterintuitive result with the fact that different units of time had been used when performing the analysis of the data: seconds for e-mails and days for letters. Furthermore the cutoff for the observation of e-mails was 80 days, while it was 30 years for the letters. If such minor details can mess up the results of the study, it is not surprising that experts question its quality. Nevertheless, Armin Bunde from the University of Giessen in Germany believes that the conclusions may not be quite as counterintuitive. The results could be explained, he says, by the fact that many more e-mails are being sent nowadays, than letters were sent years ago. In contrast to regular mail correspondence, the long waiting times for the many unimportant e-mails would therefore dominate the waiting times. Rather than supporting the study, Bunde's explanation seems all the more reason to cast doubt on the study.

All in all, there is no ringing endorsement of Barabási and Oliveira's paper. Even under the pressure of "publish or perish", scientists must do more than just produce output. It does not suffice to feed readily available data, as interesting as it may be, into a computer, let the machine crunch the numbers, and then report the smoothed, cursorily investigated results to the general public.

Chapter 13

Wobbly Tables

Presumably, everybody has been annoyed sometime in his life by a wobbly garden table. It must have been a four-legged one because three-legged tables never have this problem. Why? Because three points in space define a plane, that's why. A tabletop supported by three legs of whatever lengths, will always be 'well-defined' in space. Hence a three-legged table never wobbles, even though the tabletop may not be horizontal. A square table, unfortunately, can both wobble and stand askance at the same time. The question is whether its position can be adjusted, so that it stands firmly on the floor with all four legs.

This problem has been around since at least 1970, when the British mathematician Roger Fenn first proved a theorem which says the following: on a surface—no matter how bumpy—there always exist four points lying on a horizontal plane, which can be connected to form a square. This holds, regardless of how large the square is. In everyday language the theorem says that however uneven the floor, a square table of any size can always be made to stand firmly and horizontally. The only caveat is that the legs have to be sufficiently tall so that the tabletop rises above all the bumps. A year later, Joseph Zaks demonstrated at a conference in Louisiana that a triangular chair can always be placed on a bumpy surface in such a way that the seat is horizontal.

Both proofs have a drawback, however. They are so-called existence proofs which merely show that three resting points exist for the stool and four for the table. They say nothing about where these points are or how they can be found.

This was the starting point for the mathematicians Bill Baritompa, Rainer Löwen, Burkard Polster, and Marty Ross as well as, independently, André Martin, a physicist from Cern in Geneva. In 2005, they published articles in which they proved that not only square tables, but all rectangular tables, could be set up in such a way that they do not wobble, provided the ground is nowhere steeper than 35 degrees. They demonstrated this with the help of the so-called 'mean value theorem'. This theorem says, in effect, that if a table leg is suspended in the air above the ground in one position, and digs into the ground in another, then—during a rotation from one position to the other—there must be a point, where the leg just touches the ground. The conclusion is that a wobbly table can be stabilized by rotating it. But again there is a drawback: there is no guarantee that the tabletop will be horizontal in the wobble-free position.

The issue just described is the three-dimensional version of a two-dimensional problem, posed by the German mathematician Otto Toeplitz. It asked whether any closed curve—it can be very complicated but must not cross itself—contains four points that can be combined to a square. Toeplitz, incidentally, persecuted by the Nazis in Germany, emigrated to Jerusalem in 1939, only to die there one year later.

Toeplitz himself made a start on this question in 1911, when he addressed the problem in the Swiss Journal *Verhandlungen der Schweizerischen Naturforschenden Gesellschaft in Solothurn* (Proceedings of the Naturalist Society in Solothurn). He informed the readers that he and his students in Goettingen had been able to prove that all convex loops, that is, all unindented loops, can span at least one square. Ever since, mathematicians have been working on this problem. In 1929, the Russian mathematician Lew Schnirelman, provided a proof for convex and concave loops, i.e., loops that are dented in and dented out, provided their curvature fulfilled certain constraints.

Unfortunately, Schnirelman's proof contained an error. It was found by the mathematician and Talmud scholar Henry Guggenheimer, who completed his Ph.D. in mathematics at the ETH in Zurich, Switzerland, and then went on to teach mathematics and Jewish mysticism in Brooklyn. Guggenheimer was able to repair the error and published a corrected proof in 1965 in the *Israel Journal of Mathematics.*

The problem, however, still fascinates mathematicians and many of them are busy as ever working on different versions of it. For example: do triangles, parallelograms, rhombi, pentagons, or other polygons fit onto the loops? Even curves that do not lie on a plane but roam freely in space are being investigated. In 1991, Brian Griffiths, a British mathematician, proved that certain curves in space contain the corners of skewed squares, that is, squares with sides of equal lengths that do not lie in a plane.

Personalities

Chapter 14

Bella Abramova Subbotovskaya and the "Jewish People's University"

Exactly 25 years ago, on September 23, 1982 at about 11 o'clock at night, an accident occurred in a dark street in Moscow. A woman walked along the sidewalk. She had just visited her mother and was on her way home. It was a quiet street, hardly a vehicle passed by at this hour. Suddenly a truck appeared at high speed, hit the woman, and drove off. Moments later another car drove up, stopped for a moment next to the victim, and also drove off. An ambulance came—who had called it?—and took the victim straight to the morgue. The funeral took place the next day. It was a very low key affair, nobody talked, no eulogy was held. Mourners only whispered among themselves, all the while observed by a few official-looking men. Eventually everybody quietly dispersed. The hit-and-run driver was never found, and the case was closed. The accident had all the trappings of a KGB hit. The victim was the 44-year old mathematician Bella Abramovna Subbotovskaya. In the days preceding her death she had been summoned several times for interrogations to KGB offices. The "crime" about which she was questioned was the organization of a "Jewish People's University".

It is almost forgotten today, but not so long ago Jews were routinely denied entry to reputable institutes of higher education in the Soviet Union. Although the discriminatory practice was not limited to mathematics, it was especially glaring in this field to which Jews had been traditionally drawn. Twenty-five to thirty percent of the graduates of the high schools that were geared towards physics and mathematics were Jewish; only a handful were admitted to the top institutes. The most prestigious among them was MekhMat, the Department of Mechanics and Mathematics at Moscow State University. The driving forces behind MekhMat's adherence to the anti-Semitic admissions policy decreed from above were V. A. Sadovnichii, currently rector of Moscow University; O. B. Lupanov, MekhMat's dean from 1980 until his death in 2006; and A. S. Mishchenko, professor and senior examiner at MekhMat. But anti-Semitism in Soviet mathematics was not restricted to insignificant, small-minded people. Distinguished Soviet mathematicians were known to be pathological anti-Semites, for example, L. S. Pontryagin and I. M. Vinogradov, who wielded enormous power over the lives and careers of Soviet mathematicians but also, surprisingly, the human rights activist I. R. Shafarevich. The absurd justifications some of them gave for their virulent feelings against Jews—which were buttressed by the administrative authority some of them held—was that Jews are genetically programmed to develop mathematical abilities at an early age. By the time ethnic Russians fully develop their mathematical powers, so the reasoning went, all opportunities to study and all faculty positions are already taken by Jews. Such a situation was to be prevented by barring the latter from access to higher mathematics education right after high school. A more prosaic reason for the rabid anti-Semitism exhibited by the Soviet authorities was their cowardly desire to blame others for their economic and other failures.

During the 1970s and the 1980s, up until Perestroika, such a policy was strictly enforced. One institute to which entrance was all but barred to Jews was MekhMat. It was—and is—considered the premiere mathematical center in the then-Soviet Union and today's

Russia. Jews—or applicants with Jewish-sounding names—were singled out at the entrance exams for special treatment. Written tests, identical for all applicants, were usually no problem for gifted and well-prepared students. However, according to one source, MekhMat officials opened the written exams—which had been handed in carrying only ID-numbers and no names—identified the Jews, and drastically reduced their grades. The hurdles were raised in the oral exam. Unwanted candidates were given "killer questions" that required difficult reasoning and long computations. Some questions were impossible to solve, were stated in an ambiguous way, or had no correct answer. They were not designed to test a candidate's skill but meant to weed out "undesirables". The grueling, blatantly unfair interrogations often lasted five or six hours, even though by decree they should have been limited to three and a half. Even if a candidate's answers were correct, reasons could always be found to fail him. On one occasion a candidate was failed for answering the question "what is the definition of a circle?" with "the set of points equidistant to a given point." The correct answer, the examiner said, was "the set of *all* points equidistant to a given point." On another occasion an answer to the same question was deemed incorrect because the candidate had failed to stipulate that the distance had to be nonzero. When asked about the solutions to an equation, the answer "1 and 2" was declared wrong, the correct answer being, according to an examiner, "1 *or* 2." On a different occasion, the same examiner told another student the exact opposite: "1 *or* 2" was considered wrong. One candidate received a failing grade for making use of the "unsubstantiated inequality" $(\sqrt{6})/2 > 1$. And if an applicant, against all odds, managed to pass both the written and the oral test, he or she could always be failed on the required essay on Russian literature with the set phrase "the theme has not been sufficiently elaborated." In truly Kafkaesque manner, even a perfect score did not guarantee admittance to a Jewish student. "Grades received at entrance examinations do not play a decisive role for admission to our Institute," the prospectus of the Moscow Institute for Physics and Technology read. With very rare exceptions, appeals against negative decisions had no chance of success. At best they were ignored, at worst the applicant was chastised for showing "contempt for the examiners."

Such was the setting when, unbeknownst to each other, two courageous individuals, Valery Senderov and Bella Subbotovskaya, decided to do something about the sorry situation. Senderov, who had done work in functional analysis, was a mathematics teacher at Moscow's famed "School Number 2", and Bella, who had already published important papers on mathematical logic, held positions at various technical research institutes performing programming tasks and numerical computations. The two met by coincidence in July 1978 on the steps of the main building of Moscow State University, where the entrance exams to MekhMat were taking place. Their aim was to assist students who had just failed the oral exams with the formulation of letters to the Appeal Committee. Senderov had a further aim in mind: together with his colleague Boris Kanevsky, he was going to document the racially motivated bias and unfairness in the MekhMat entrance exams. Senderov was just talking to one of the flunked students when the examiner rushed out and challenged him. An altercation ensued that soon degenerated into a scuffle; security was called, and Senderov was forcefully removed from the premises. This event marked—as Kanevsky recounted at a recent memorial session at the Technion in Haifa in tribute to Bella—the beginning of an ambitious and dangerous undertaking, the creation of a "Jewish People's University."

Bella Abramovna is described by her friends and admirers variously as loud, energetic, and demanding, but also as warm, kindhearted, optimistic, with great courage and resolve. She had fallen in love with mathematics beginning in first grade and that love never abated, even though she also informally prepared for a career in music and played several instruments. As an educator "she had the ability to convey her perception to the most varied types of people," her husband Ilya Muchnik would later write. She could evoke appreciation for her subject in almost all persons with whom she dealt, be they grade-school children for whom she designed mathematical games; adults attending evening school, weary from a full day's work; or gifted high-school graduates who were denied entry to Moscow State University.

Bella and Ilya met at a seminar on cybernetics where a paper on how to compose music on a computer was discussed. Bella, who had studied violin for ten years at the music school, and Ilya, who had the idea of studying the statistics of musical fragments in Jewish folk songs by computer, immediately took a liking to each other. After about a year, in the summer of 1961, they decided to get married and moved into a six square meter room with a stove-heater and an outhouse in the yard. They lived in poor surroundings in a beehive of buildings, each of which was occupied by three or four Jewish families, complete with numerous children and grandparents. The common language among the neighbors was Yiddish. The wedding was a very authentic affair held in the yard, with everybody singing Jewish folk songs, accompanied by Bella on the violin.

After their marriage, Bella made a meager living performing engineering tasks for various technical research institutes. She did not like her routine work but did it diligently nevertheless. A change was brought about when the couple's daughter began studying at high school. Bella started wondering where children of her daughter's school, many of whom were Jewish, would pursue their studies after graduating. This is when she became painfully aware of the dead-end that awaited Jewish children. Even the most gifted among them had practically no hope of studying at first-rate institutes. Bella herself had been lucky enough to attend MekhMat in the mid-1950s, a period after Stalin's death and at the beginning of the Khrushchev era, when Jews were not yet discriminated against. But by now, in the late 1970s, the situation had vastly deteriorated. Bella decided to devote herself to furthering the ambition of dedicated and mathematically gifted high-school graduates. She helped prepare them for the entrance exam to the faculty of mathematics and assisted those denied entry in writing the necessary letters to the appeals committees.

Meanwhile Senderov and Kanevsky wrote the underground classic "Intellectual Genocide" in which they documented the results of their investigations of failed Jewish MekhMat candidates. The mathematical economist Victor Polterovich had collected statistics on the admission of students from Moscow's leading mathematics and physics high schools to MekhMat. In 1979, of the 47 non-Jewish students who applied, 40 were admitted, but only six of the 40 students with Jewish names. This was after a kind of self-selection had already taken place, with many Jewish students not even applying. The questions given to candidates with Jewish names were distressingly difficult, and the reasons for failing the students or denying their appeals were equally hair-raising. Polterovich also wrote a "Memo for students applying to MekhMat who are thought of as being Jewish", which was distributed by Senderov and Kanevsky. But then Bella did much more. She decided to partially restore hope and fairness by giving the rebuffed students an opportunity to obtain a fundamental mathematical education at her home.

Since appeals to the appropriate committee were of no avail, the failed students were left with no option but to study at institutions that prepared them for professional careers, like the Institute of Metallurgy, the Pedagogical Institute, the Institute of Railway Engineers, or the Institute for the Petrochemical and Natural Gas Industry. They would get a solid grounding in applied mathematics but would have no hope of ever glimpsing beyond the immediate areas of the professions for which they were trained. Pure mathematics would remain out of reach.

But Bella would have none of that. In the fall of 1978 she started an ambitious and unprecedented undertaking in her own home: the "Jewish People's University." Bella's former classmate Alexandre Vinogradov, who had received his doctorate from MekhMat fifteen years earlier and was now a professor at that institute, devised a nonstandard advanced study program for the initial course. Together with former and current Ph.D. students, he taught the initial course. (Because of ideological differences with other faculty members, Vinogradov left the project after a few months. The point of contention was whether Bella's university should limit itself to teaching mathematics or be part of the broader struggle against the Soviet regime.) The university began as a study group with a dozen or so students, but news about the undertaking quickly spread by word of mouth. No equipment was available except for a children's chalkboard standing on an unstable tripod. Later, a more suitable blackboard was obtained. Since it could not fit through the narrow staircase of the tenement where Bella (now divorced) lived, it had to be hoisted through the fifth-floor window. Bella was the guiding spirit behind every aspect of the unique undertaking. She herself did not teach, but solicited the help of former classmates, now established mathematicians, to lecture at her university. The informal institution was open to everyone, but most students and many teachers, though by no means all, were Jewish.

And there was no lack of gifted teachers; the recruited faculty was of the highest caliber. The courses taught in Bella's apartment, and later at other venues, corresponded to the first two years of the MekhMat undergraduate curriculum. Vinogradov, Senderov, Alexander Shen, and Andrei Zelevinsky taught calculus; Dmitry Fuchs differential geometry and linear algebra; Alexey Sossinski, a Russian born in Paris and brought up in America, lectured on modern algebra; Boris Feigin gave courses on topology and commutative algebra; Victor A. Ginzburg taught linear algebra; Mikhail Marinov—who, after having applied for an exit visa to Israel, labored as a construction worker—lectured on quantum mechanics and field theory; seminars were run by Boris Kanevsky. Altogether 21 people taught at the university. Universities all over the world would have been proud to have a faculty of the quality found at Bella's Jewish People's University. Nobody received any money. The teachers took on the selfless and risky task motivated solely by human decency, to right a wrong, and out of love for mathematics. There was even a "visiting" professor: once, during a trip to Moscow, John Milnor came to lecture.

Word of the underground university got around, and the student body grew. Soon, the auditors no longer fit into Bella's minuscule apartment. Other venues were sought and used—with and without permission: classrooms in elementary schools, empty study halls in the university's law department, the chemistry building, the humanities building, the Institute for the Petrochemical and Natural Gas Industry. In 1979, the second year of the Jewish People's University's operation, about 90 students attended its classes. Bella did everything, organizing the meetings, calling the students to inform them of the schedule and venue, even distributing tea and homemade sandwiches during the breaks between the lectures. One important and risky undertaking that she organized was the samizdat preparation and distribution of lecture notes. At first they were typed and re-typed using carbon copies, equations being inserted by hand. Eventually they were photocopied. Nobody dared ask how and where, since unauthorized duplication was considered a serious crime in the Soviet Union. In 1980, study sessions were increased to twice a week. Saturdays were reserved for three lectures and a seminar.

Even though some of the faculty members and students, especially Senderov, were known dissenters of the Soviet system, any mention of politics was carefully avoided by the teachers at Bella's university. But the enterprise was becoming too successful for the authorities to ignore. Even though it had no political intent whatsoever, it defied the Soviet system on a grand scale. The authorities could not allow an unofficial and independent institution to flourish, thereby challenging its sole claim to authority. The mere existence of the Jewish People's University was considered a political act of resistance by the authorities. The end loomed near.

At the beginning of the university's fifth year of operation, Bella was summoned to KGB offices and interrogated. It had been known all along that KGB agents had attended lectures in order to observe the goings-on. They must have known that no subversive activities were carried out at the Jewish People's University. But they never comprehended what kind of institution Bella's university was. The agents just could not grasp that people were willing to teach mathematics without being paid. One day in the summer of 1982, news came that Senderov, Kanevsky, and a student, Ilya Geltzer, had been arrested. Another young man, Vladimir Gershuni, was arrested together with them and later forcibly confined to a mental institution. They had distributed leaflets protesting unpaid "volunteer" work that the Communist Party demanded of loyal citizens on the Saturday commemorating Lenin's birthday. Senderov and Kanevsky were known dissenters of the Soviet regime but had always kept mathematics and politics strictly separate. Nevertheless, their and the student's affiliation with Bella's enterprise gave the authorities the excuse they sought.

Bella was summoned again and asked to serve as a witness against Senderov. Of course, she refused. Her independent spirit would not allow anything but defiance of authority. The tragic consequences occurred a few days later. The bus of Moscow Sate University's chamber orchestra, where she played first viola since her student days, took her body to the cemetery. Her ashes were later buried at the Jewish cemetery Vostryakovo.

Bella's death spelled the end of the Jewish People's University. Senderov was sentenced on charges of anti-Soviet agitation and propaganda to seven years in prison—where he would spend long stretches in punishment cells sustained by a meager diet that left him too weak to even rise from his bunk. He was released by Gorbachev after Perestroika, having spent five years in prison. Kanevsky was sentenced to one year and two months in prison. Seminars continued for a few more months due to the valiant efforts of some remaining faculty members, but without Bella's support and guiding hand the spirit was missing. In the spring of 1983 the institution finally closed its non-existent doors. During the four years of its operation, the "Jewish People's University" had instructed about 350 alumni in higher mathematics and brought forth about 100 "graduates," some of whom would become professional mathematicians and faculty members at prestigious institutions, mostly in the United States and in Israel. But Bella had given her alumni more than just a math education: in the face of injustice, discrimination, and seemingly insurmountable difficulties, she had offered them hope and taught them to fight back.

Chapter 15

No Answer from Professor Ekhad

It is for good reason that professors of mathematics are sometimes thought of as being bizarre. Take, for example, Professor Shalosh B. Ekhad from Rutgers University. A look at his list of publications leaves you suitably impressed. During the past ten years alone, he published dozens of articles, sometimes in collaboration with other authors. One of his first achievements was the proof of the so-called cosmological theorem, a tricky conjecture for which John H. Conway, Professor of Mathematics at Princeton University, claimed to have already had a proof. Unfortunately, he subsequently lost it before anybody was able to subject it to verification. So Ekhad put himself to it and before long came up with a completely new proof of his own. This put Ekhad's name firmly on the mathematicians' map. Ever since, colleagues around the world do not cease to quote from his publications, mainly from his proofs in combinatorics.

Something was strange, however, with Ekhad and nobody could quite figure out what it was. His name could not be found on any university's faculty list and the return address noted at the bottom of his publications did not correspond to any actual office at Rutgers University. Invitations to seminars or for guest lectures either remained unacknowledged or were answered, always in the negative, by a polite secretary. Requests by potential graduate students who wished to write their doctoral theses under Professor Ekhad's supervision were routinely declined.

Who could this timid mathematician be who so shunned the public? Readers who might be familiar with the Hebrew language will notice that Shalosh B. Ekhad stands for "three in one," a symbol for the trinity. Soon, rumors spread over the Internet proclaiming that behind this mysterious mathematician hid a missionary who was intent on converting the world. But how could mass conversion be accomplished with the help of esoteric papers on combinatorial theorems? While our book here deals with mathematicians and not with detective work, this was far too intriguing a puzzle to remain unsolved. When "yours truly" heard that one of Ekhad's co-authors, Doron Zeilberger, also from Rutgers University, would attend a workshop on the Greek island of Mykonos, he immediately sprang into action. Sparing no costs, he caught the next available flight to the sunny island in the Mediterranean Sea. The event turned out to be one of these congenial professional seminars, with experts from all corners of the world convening to exchange ideas. This author's hope was to cozy up to Zeilberger and eventually zero in on Ekhad.

At first Zeilberger tried to camouflage himself, dressed down in shorts and sandals, as is appropriate for a congress in a Greek vacation spot. But the name tag pinned on his T-shirt quickly gave him away. His cover blown, he quickly spilled the beans. Nobody by the name of Shalosh B. Ekhad existed! The mysterious author of advanced papers in combinatorics was none other than a computer.

In 1987, Zeilberger's first personal computer was a machine manufactured by AT&T. The model had been developed at Bell Labs, in building No. 3, Corridor B, room number 1. So, naturally, it was marked as 3B1. Such was Zeilberger's pride in his new toy, that he promptly gave it a name in his mother tongue: Shalosh B. Ekhad.

For Zeilberger, 3B1 was more than a toy; it soon became his companion and friend, whom he intended to teach nothing less than how to find and prove mathematical identities. Together with Herbert Wilf from the University of Pennsylvania, Zeilberger developed an algorithm which allowed computers to do exactly that. Their joint work was awarded the 1998 Leroy P. Steele Prize of the American Mathematical Society.

Shalosh B. Ekhad soon surpassed even Zeilberger's already high expectations. All that was required was to input a few initial instructions and off it went humming and buzzing happily for a few hours, or days, before spitting out results. "Shalosh found new proofs for already-known identities and also came up with some completely new identities," Zeilberger explained with obvious paternal pride. For mathematicians, who live and breathe their subject, some results appear to be beautiful, others less so. But whatever their perceived degree of beauty, many were very useful. "Still others were neither beautiful nor useful," Zeilberger noted dryly, "and they could simply be ignored."

Ekhad's godfather has high ambitions for his scion. He forecasts nothing less than a paradigm change. In the future, computers will discover fundamental relationships, he believes, thus, far surpassing human capabilities. "It will only be a matter of a few decades until more and more original mathematical research will be conducted solely by computers," says Zeilberger, who clearly enjoys being provocative. Then, theorems that were the pride of many a human mathematician back in the twenty-first century will seem like Mickey Mouse mathematics. But, Zeilberger concedes, this trend will also tremendously increase the importance of teaching. So, there clearly still is something to look forward to for the nonsilicon mathematicians.

Chapter 16

The Yippie Mathematician

Stephen Smale is one of the most prolific mathematicians of our time and certainly one of the most dazzling. Recipient of a Fields Medal in 1966, he was also known for his anti-Vietnam war protests and the co-founding of the Yippie movement in the late 1960s. At one time he was even subpoenaed by the House Committee on Un-American Activities. He also ran into trouble with the National Science Foundation when he declared publicly that he had done some of his best work on the beaches of Rio.

In May 2007 he was awarded, together with Hillel Furstenberg, the Wolf Prize in mathematics. In Jerusalem for the prize ceremony, Smale granted an interview in which he talked about his work and career.

—Andy Magid

Professor Smale, why is mathematics important to you?

Oh, that's a tough question. Maybe I'm different from other mathematicians. I consider it as just one important thing to study. I see myself broader, as a scientist, even a little bit of an artist. So mathematics is not the sole motivating thing in my life, far from it. But I do see a beauty in mathematics because of its elegance and its ability to idealize the things you see in everyday life. Understanding the things around you has been the motivating factor for me for the past 40 years.

So is mathematics a cultural endeavor?

Well, I wouldn't say so much cultural, as science in a broad sense. The traditional motivation to do mathematics is to understand physics, but also to understand, say, economic phenomena. I am now trying to understand human vision, trying to develop something like a model for the visual cortex. Maybe it will turn out that there are some universal laws and eventually we will understand how humans learn and think. That's an example of how mathematics can help us understand natural phenomena.

Why is mathematics so effective in explaining phenomena, as opposed to, say, narratives?

Mathematics is a kind of formalized way of thinking. One can be much more precise in mathematics than in literature, express relationships in a more precise way, include magnitudes. And even fuzziness can be incorporated in mathematics by using probabilities. I use that a lot because when moving from physics to vision and biology one has to incorporate some kind of fuzziness. The way I do that is—in the mathematical tradition—by using probability.

Mathematics is so effective because one can look for universal laws more easily with mathematics than without. It enables us to abstract the main ideas. With formalization and symbols one is able to see what is universal. The abstraction allows us to see universal ideas. I have been very inspired by Newton who could see a falling apple and the motion of planets and recognize them as part of the same phenomenon. I would like to see a language that allows us to translate what we see and then recognize it as part of a broad phenomenon.

Kepler's Conjecture was believed to be correct, even before it was proven, and many people believe the Riemann Conjecture to be correct. Why is it so important in mathematics to be rigorous when proving something?

Just because a lot of people believe something does not mean it is true. I am in favor of rigorously proving big problems. On the other hand, I am not quite so devoted to the idea that proof is the most essential thing in mathematics. What may be more important are the relationships of the main structures, the concepts, and the development of these concepts. Proofs are often an important part of that but are not the main focus of my work. I'm rigorous, I try to have things correct, but sometimes proofs are almost secondary to seeing how the main structures are laid out. I look at relationships between mathematics and eventually between parts of the real world.

Do you accept computer proofs?

Since proofs are not the ultimate in mathematics for me, computer proofs are okay. Maybe not as good as a construction, a structured conceptual proof, but okay.

Your career covered four areas: topology, dynamical systems, mathematical economics, and computer science. Why did you leave topology?

In 1961 I did change subjects. I did not change completely, but I did leave topology. I said it publicly. I had proved Poincaré's conjecture in dimensions five and greater and I thought, after that, things were a little anti-climactic. Proofs for the third and fourth dimensions were still missing but it seemed—I won't say I was right—that these were just special cases. So it was more exciting for me to understand the dynamics of a discrete transformation of the two-sphere than working out Poincaré's conjecture.

Were you convinced at the time that Poincaré's conjecture was correct?

Oh no, far from it. I even had a counter-example. But it did not work, I found a mistake. Whenever I work on a mathematical problem I work on both sides of the question because they reinforce each other. If you work only on one side you don't get such a good perspective. One should not have too many pre-conceived ideas. Sometimes you should say "well, if it's not true, how would you go about proving that?" Going back and forth is an important part of proving a theorem.

What did you do after topology?

I had been doing dynamics for some years before that and had some idea about the great problems in dynamics. So I started working on those problems. Then I also did some work in electrical circuit theory, in physics, in mechanics.

How did you get into economics?

Well I was always interested in economics because of my political activities and my contacts with so many Marxists. One day Gérard Debreu, who later received the Nobel Prize for economics, came to me and asked me some mathematical questions about equilibria, and I told him about Sard's theorem, which was relevant to his research. A friendship grew between us. I learned a lot from him and he from me. We never worked together but we talked a lot. In fact, I helped him get the Nobel Prize. Ken Arrow and I nominated him to the Nobel committee.

Then you left economics and got into the field of algorithms.

Yes, after a few years. I had developed algorithms to find economic equilibria. I was not trying to simulate. I was just trying to find an abstract mathematical algorithm; other people simulated it. Given supply and demand, the task was to find the equilibrium prices in the economy. And I was doing it in the general setting of a possible economy. There was another algorithm by Herbert Scarf. I believe mine was faster and more natural. This led to the question, Which was better? So I continued towards computer science in order to understand why one algorithm is better than another.

Was your algorithm supposed to describe the workings of the economy?

No, not really. There are a couple of issues here. One of them is, how the economy works, how prices adjust. For me that was the biggest unsolved problem in economics. I spent time on that and failed. There is another problem. If parameters change, how do economic agents find the changing equilibrium? How do they locate it numerically? And I did the theory, the algorithm, of how to do that.

Was it your aim to aid a centralized economy find the equilibrium?

I was never quite so strong on that. As a student, before I was a Vietnam war protester, I was also a communist, but never because of the economies of Vietnam or Russia. I did not know that much about economics, and I was already somewhat disenchanted by that. As I got older I abandoned Marxism. But it took many years, experience of the world, maturing intellectually, and seeing what was happening. I was never interested in Marxism from the point of view of a planned economy.

Later, I became interested in understanding markets. But I am not a believer in the capitalist system, far from it. Let's say that over the years I became market-oriented. So when I got into algorithms, I was inspired by the market economy; on the assumption that the market gives us equilibria, how does one find them? They are given by equations, and I was providing algorithms for solving these equations.

What are you working on now?

While in Israel [for the Wolf Prize ceremony] I will give three talks. At the Weizman Institute I will talk on the mathematics of vision. It is some kind of visual cortex model, but more universal. Then I go to Haifa University where I will speak on data, the geometry of data. One sees all these data points and wants to find an underlying geometry. So I am going back to topology a little to do that. Data is the main thing one is trying to understand, and I am looking at the geometry, or the topology, of data. It is not quite pattern recognition; my first talk is connected to pattern recognition. All these things are a little bit mixed together but they are different.

Then I will talk in Beersheva on the flocking of birds. It's a big thing in zoology, there are lots of observational studies. You have a bunch of birds on the ground, and then they suddenly all go up in the air and fly together at the same speed. It has to do with control theory, and robotics people want tiny robots to communicate with each other. They are the same phenomena. So they want to see how they can organize these kinds of common phenomena. A similar phenomenon occurs when a language emerges. The idea is how one can reach common understanding through the senses. In economics it would be the belief in a common price system, a necessary condition for prices to operate. So it goes back to my old question in economics: How do people arrive at a common belief in a price system?

You have observed mathematics for half a century. Where do you think the field is going?

My feeling is that there is a shift in mathematics away from traditional areas of physics. It used to be a big area for mathematics and for thousands of years inspired a lot of mathematics. But mathematicians seem focused too much on physics. I believe that things are changing much more in mathematics than in physics. Like the areas that I work in, like vision and the other questions coming in from biology, statistics, engineering, computer science, and especially computation. A lot of these things influence the way that mathematics is changing. So where is mathematics going? It is leaving physics to a great extent and moving into the areas I just mentioned.

These are areas of applied mathematics. What about pure mathematics?

I am not talking about applied mathematics. I don't believe in that dichotomy. I am talking about using mathematics to understand the world. When developing calculus and differential equations Newton was doing mathematics in order to understand the laws of gravitation. Did he do applied math? I don't think so. Did he do pure math? No. So that's the kind of mathematics I am thinking of. It's not what it was 150 years ago. Problems come down more from computer science, engineering, and biology. But it's mathematics proper, it's not applications.

Chapter 17

Sibling Rivalry

On 16 August, 2005, the city of Basel in Switzerland commemorated the 300th anniversary of the death of Jacob Bernoulli, one of the world's most famous mathematicians. Son of a spice merchant who also held the prominent role of town councillor and judge, Jacob was the eldest of this notable family. During the seventeenth and eighteenth century it could pride itself in counting no fewer than eight world-renowned mathematicians amongst its members. Although young Jacob showed a keen interest in mathematics early on, he at first followed his father's wishes and aimed for the ministry, dutifully studying theology. This did not prevent him, however, from secretly pursuing the two real loves of his life: mathematics and astronomy.

Being a young and educated man, his services as a private tutor were eagerly sought in Geneva where Jacob lived at the time. Well-to-do families considered themselves fortunate if they were able to engage him for their children. After his stint at teaching, Jacob embarked on travels throughout Europe as was the custom at the time. He visited France, Holland, and England, always making sure to establish good contacts with famous scientists. Upon his return home, Bernoulli taught mechanics at the University of Basel and devoted his spare afternoons writing mathematical papers.

It did not take long for his scientific works to be recognised and after five years he was appointed Professor of Mathematics in Basel, remaining in this position for the rest of his life. His most famous student was none other than his younger brother Johann, who was quite his equal in talent. He, too, had at first pursued another career—in his case, medicine—according to the wishes of his father. But Johann, like Jacob, persevered in his mathematical interests and luckily found a secret tutor in his older brother Jacob. Together they further developed the then innovative differential and integral calculus invented only a few years earlier by Newton and Leibniz.

Unfortunately, brotherly love soon gave way to spiteful hatred which attracted nearly as much public attention as did their mathematical breakthroughs. When Johann went so far as to show off his own results while publicly denigrating those of his older brother, Jacob hit back. He announced that Johann actually had done no more than simply reproduce results which he, Jacob, had taught him previously. Johann riposted, and so it went back and forth. This part of the family history makes for disappointing reading, especially since these two towering personalities, each in his own right, belonged to the elite of mathematicians, not just of that era but of all time. While Jacob quite obviously suffered from feelings of inferiority, both brothers had an exaggerated craving for recognition. But who knows, maybe it was precisely the rivalry between them which spurred the two brothers to ever higher achievements.

Jacob's most important work was *Ars Conjectandi* (The Art of Conjecturing). Published in 1713, eight years after his death, it heralded the era of statistics and probability theory. As early as 1689, Jacob Bernoulli had already introduced the "law of large numbers," which basically says that the probability of a phenomenon is equal to the frequency of its appearance in many repeated experiments. With this, Bernoulli defined probability for the first time as a number between zero and one, thus providing the mathematical basis for a concept that heretofore had mainly been a synonym for philosophical or legal arguments.

The main sections of his book had been developed by Jacob back in the 1690s but he had never found enough time to finish it. At the time of his death only the first three parts of his work were complete. They included his theory of combinatorics and applications of the concept of probability to games of chance. The fourth part, which should have applied the theory to legal, political and economic decision making had, unfortunately, been left incomplete. For years, German and French scientists urged Johann to complete and publish his older brother's manuscript. But sibling rivalry lived on beyond the grave. Jacob's widow and son remained deeply suspicious of Johann and did not allow him near the manuscript. Johann, on his part, let it be known that he had better things to do than edit the work of a brother for whom he had never particularly cared. But eventually the expectation of even greater fame for her dead husband got the better of Johann's widow. Begrudgingly, she allowed Nicolaus Bernoulli, the son of a third brother, who had been Jacob's secretary, access to the main body of her husband's work. At long last, this part of Johann's manuscript got to see the light of day and the world was the richer for it.

Chapter 18

A Diplomat With a Love for Numbers

Although many lay people like to think that mathematics is a secret book which ought to remain under lock and key, there are others who find any odd conundrum too attractive a challenge to resist. Such was the fate of Fermat's Last Theorem. For centuries, solving it became the ultimate challenge for a huge number of both professional and amateur mathematicians. It was only in 1994 that Andrew Wiles, Professor at Princeton University, was able to solve the problem. He proved that no whole numbers a, b, and c exist such that $a^n + b^n = c^n$, for any n greater than 2.

Fermat's Last Theorem is a so-called Diophantine equation. These equations allow only whole numbers, so-called integers, as solutions. Fermat's equation is only one among innumerable equations of this kind. Carl Friedrich Gauss, a German mathematician of the nineteenth century, made no bones about the fact that Fermat's Last Theorem held only limited interest for him. He could easily jot down a whole list of similar equations, he said, for which it would be difficult to determine whether integer solutions exist or not. His successor at the University of Göttingen, the mathematician, David Hilbert, shared this view. Voicing a distinct lack of enthusiasm, he stated that Fermat's Last Theorem was merely a "special and apparently unimportant problem."

There are Diophantine problems where integers are sought that solve several equations simultaneously. An example is the search for six whole numbers a, b, c, d, e, and f for which the equation $a^k + b^k + c^k = d^k + e^k + f^k$ holds true for k equal to 2, 3, and 4 simultaneously.

In 1951 an Italian mathematician took a step, albeit not a big one, toward the solution of this problem by demonstrating that not all six numbers can be positive. Then, nothing further was heard about this problem for about half a century until one solution was discovered in 2001. (All that will be revealed here is that a equals 358 and b equals -815.) The question remained whether this was the only solution to the system of equations or whether there are also others. Another three years passed until this question was answered in 2004: not only does there exist more than one solution, there are, in fact, an infinite number of them.

Surprisingly, Ajai Choudhry, the author of the proof, published in the *Bulletin of the London Mathematical Society*, was not a professional mathematician. He was the Indian Ambassador to the Sultanate of Brunei. Born in 1953 in Uttar Pradesh, a state in northern India, Choudhry was known to be brilliant at math at a very young age. A career in mathematics seemed the obvious choice for young Ajai. But after graduating from university with flying colours in all his subjects, Choudhry, then 23 years old, decided to give academia a pass and devoted himself to the diplomatic services. For the following eleven years, the promising young man devoted himself heart and soul to India's Foreign Ministry. His duties would demand that he travel far and wide and, inevitably, mathematics faded into a childhood memory. After serving in Delhi for a while, Choudhry was transferred to Kuala Lumpur, Warsaw, Singapore, Lebanon, and Brunei.

At a diplomatic gathering at one of the embassies in Warsaw, the Indian diplomat chanced upon Andrzei Schinzel, a number theorist from the Polish Academy of Sciences. Only a few minutes into the conversation, Choudhry felt his old passion resurge and his fire for mathematics was rekindled. Thankfully, diplomatic service left Choudhry sufficient free time to devote to Diophantine equations.

Not one to do things by halves, Choudry published no less than 45 papers in scientific journals in the course of the following years. He also scooped up a prize for the proof of a theorem about the seventh powers of integer numbers. But even the heady combination of diplomacy and mathematics seemed too little to satisfy the diplomat's intellectual ambitions. He spent his leisure time poring over the chess board. Here, too, Choudhry rose to the top, becoming an International Master. In 1998 he was one of 30 opponents chosen to play in a simultaneous exhibition tournament against Anatoli Karpov, the then-reigning world champion. Choudhry's game ended in a draw for the simple reason that he had to leave the chess board in order to attend to an urgent diplomatic matter.

Purists would deem Choudhry's proof of the existence of an infinite number of solutions as somewhat trivial. In fact, although far from simple, it does not require any deep knowledge of mathematics. With the help of some transformations, Choudhry first established a relationship to so-called elliptic functions which had already played a significant part in Fermat's theorem. And from that point on, it was a mere hop, skip, and jump to reach the conclusion that an infinite number of solutions exist. It becomes clear that these solutions are by no means simple just by looking at one of Choudhry's examples. Choudry mentions a case where a equals—230,043,367,232,999,423.

Chapter 19

The "Wunderkind"

In February 2010 the mathematical world marked the 155th anniversary of the death of Carl Friedrich Gauss (1777–1855). As a toddler, the man who would later become known as the prince of mathematics, already knew enough maths to correct an error that his father made when calculating the salaries of his employees. He went on to amaze his primary school teacher with his obvious gift for mathematics and is today rightly considered one of the most important mathematician of all times.

It was only by sheer luck, however, that his genius was recognized and nurtured. In nineteenth century Germany it was quite normal for teachers to be in charge of classes comprising as many as 50 or 60 children, all of them of different ages, talents and levels of achievement. For the teaching staff to keep such a large group of kids in check, let alone teach them, was a challenge in itself. For the overstretched instructors to also be able to pinpoint the particularly gifted among them must then be seen as a true credit to them. The fact that Gauss's teachers did, in fact, recognize the young boy's early genius and that they brought it to the attention of the Duke of Braunschweig-Wolfenbuettel, deserves, therefore, special mention.

The Duke awarded Gauss a fellowship which allowed him to study in Braunschweig, Helmstedt, and Göttingen. Gauss's first breakthrough in mathematics occurred when the seventeen-year old youth discovered that a heptadecagon—a regular seventeen-edged figure—could be constructed by compass and straightedge. The problem had puzzled mathematicians since the days of the Ancient Greeks. So pleased was Gauss with his result that he resolved to make mathematics his career, rather than continue to study philology which was another love of his life.

In 1807 Gauss was appointed the first director of Göttingen's observatory. Even when his reputation was already established throughout Europe, he never considered accepting lucrative professorships elsewhere. Remaining loyal to Göttingen, he continued to hold his post until his death. As for his personal life, it was, sadly, overshadowed early on by tragedies and loss. After giving birth to their third child, Gauss's beloved wife Johanna passed away, soon to be followed by their daughter and then by Louis, the baby boy himself. The deaths left Gauss distraught and he plunged into a deep depression from which he would never fully recover. Many years later, as an old man, Gauss wrote in a letter to a friend that *"it is true, my life had much for which the world could envy me. But, believe me, the bitter experiences ... outweigh whatever was good more than a hundred-fold."*

The significant contributions Gauss made to mathematics—number theory, statistics, analysis, differential geometry, probability theory, and other sub-fields—are too numerous to list. Suffice it to say that the Encyclopedia of Mathematics mentions his name no fewer than 485 times. Add to this the adjective "Gaussian" and you will find the even higher number of 1,370 references to the famous mathematician. But Gauss was not only a mathematician. He was also an expert in astronomy, physics, geodesy, optics, and electrostatics. Together with his colleague, the physics professor Wilhelm Weber, he constructed the first electromagnetic telegraph in 1833. It connected Gauss's study in the observatory to Weber's office, which was one kilometer away.

Just how significant Gauss's impact was on the development of modern science can be judged by considering just one of his many outstanding achievements. It would become an indispensable component of one of the most important advances of the twentieth century: the development of the general theory of relativity.

How did that come about?

In 1828 Gauss was commissioned to carry out a survey of the Kingdom of Hannover. This was a purely mechanical task, quite beneath someone of Gauss's stature. But Gauss would not have been the outstanding mathematician he was had he not seized the opportunity to make several important advances in the mathematics of curved surfaces. Braving bad weather, unhelpful assistants, malfunctioning instruments, and countless nights in dingy inns, Gauss devoted the summers of the next two decades to produce as accurate a map of Hannover as was possible. The result was quite extraordinary—and it is not the map of Hannover that I mean here. Gauss had done nothing less than develop the framework for an entirely new discipline of mathematics: differential geometry.

For a long time, Gauss had carried with him the suspicion that geometry based on Euclid's five axioms was not the one and only truth. The axioms can be paraphrased as stating that any two points can be connected by a straight line and that each straight line has parallels. While these propositions make sense in the flat plane, they do not hold true on a curved surface, such as the earth, which is where Gauss, after all, performed his measurements. The following example may illustrate this: The shortest line between the Statue of Liberty in New York City and the Eiffel Tower in Paris lies along a so-called great circle. Now, take a location exactly ten miles north of the Statue of Liberty and another one exactly ten miles north of the Eiffel Tower. The shortest line between these two points again lies along a great circle. But the two paths are not parallel. Extending them and following them along the two great circles, they cross twice: once somewhere off the coast of Somalia and once more somewhere in the Pacific Ocean.

Gauss called the shortest path between two points on a curved surface—corresponding to the straight line in a plane—the geodesic line. A student of his by the name of Bernhard Riemann built upon this notion to further develop differential geometry. The new discipline carried great mathematical interest, but had little practical significance. But in the early twentieth century Albert Einstein seized upon it. After formulating the theory of special relativity in 1905, Einstein spent the following decade intensively studying differential geometry. The preparatory work accomplished by Gauss and Riemann provided Einstein with the necessary tools to formulate the theory of general relativity in 1916.

In 1687 Isaac Newton, the founder of modern physics, had formulated his laws of motion. The first one said that if no forces act on an object, this object will either remain motionless in space or will continue moving along a straight line. Einstein refined this law by replacing the straight lines with geodesic lines. And since space is curved due to the existence of mass and energy, as Einstein envisaged in his thought experiments, an object which is neither pushed nor pulled moves along a geodesic line in four-dimensional space-time.

Chapter 20

Brilliant but Fallible

Daniel Biss was a brilliant student of mathematics. After graduating *summa cum laude* from Harvard at the tender age of twenty, he swiftly moved to MIT to obtain a Ph.D. A research scholarship from the Clay Foundation followed straight on. His early claim to fame was established by two publications in particular. In 2003 his landmark papers on so-called "Grassmannian manifolds" appeared in the highly acclaimed *Annals of Mathematics* and *Advances in Mathematics*.

A brilliant career seemed to be in the offing, but suddenly, the news came that this promising mathematician had left academia and turned to politics. The former assistant professor embarked on a campaign to become the Democrat's State Representative in Illinois' 17th District. "I felt that I could contribute more to society by getting involved in politics than in mathematics," Biss explained later.

Truth be told, by then, Biss's mathematical star had already lost a lot of its luster. Nikolai Mnev, a mathematician at the Steklov Institute in St. Petersburg, Russia, had taken a closer look at Biss's proofs and detected a subtle error. In response to Mnev's complaint, Biss wrote that he had already been alerted to this problem by other experts in the field and that it would soon be fixed. Laura Anderson from State University of New York at Binghamton was in the process of ironing out the problem.

Years passed, however, without Anderson, Biss, or anybody else, for that matter, publishing a correction of the mistake. On the other hand, Biss also failed to withdraw his work. Mnev received no clear answers to his repeated requests for clarification as to where things stood. Robert MacPherson from the Institute for Advanced Study in Princeton who was editor of the *Annals of Mathematics* when Biss's work was published, also chose to remain silent. Mnev was irritated and let it be known. Biss was a nice chap, he wrote to friends; his advisors were eminent mathematicians; the journals in which the proofs were published were serious. But apparently, the system could simply not cope when an unprecedented event occurred.

In September 2007, Mnev, by then totally frustrated, put a two-page note onto the Internet. First, he expressed his regret at having to draw attention to the serious flaw, but he regarded it as his unfortunate duty to do so. While the mistake he had uncovered rendered Biss's work invalid, it still had not been withdrawn after four years. This was even more alarming, Mnev felt, since other mathematicians, who were not yet aware of the mistake, had, in the meantime, begun to build upon the faulty proof. Soon, much effort would have been spent in vain on developing theories that were based on an erroneous result.

It took yet another year for Biss to finally admit that the proof of his main result was not fixable. On 11 November 2008 he finally submitted an erratum to the *Annals of Mathematics* and one month later to the *Advances in Mathematics*. In between these letters there was more bad fortune for Bliss: he narrowly lost the election in his district by 1,774 votes.

This should have been the end of a sorry chain of events. But the journals took their time in publishing the errata. While it had been very painful to the young author to admit his mistake, the journals apparently found it more painful still to own up to their fallibility. Only in March, 2009, four months after Biss had sent his letters, and only after outsiders began demanding explanations, did the journals recall the flawed pieces of work from their websites.

Thankfully, there is always somebody who sees the bright side of things. In this case, it was Robert MacPherson, the previous editor of the *Annals of Mathematics*. "It is extraordinary," he commented, "how mathematics tends to correct itself." Mistakes cannot persist; sooner or later they will be found out. And what does Mnev—who for years had unsuccessfully tried to have the flawed papers corrected—think about the whole affair? It reminded him of the communist era, he writes with a wry sense of humor. If you wanted a leaky roof repaired, you were better off contacting a reporter from the party newspaper *Pravda*, than getting officials to have the repair done.

In the air

Chapter 21

The Plane Facts

Obviously, airlines only make a profit when their planes are airborne. It seems quite logical, therefore, that turnaround times, the periods the aircraft spends on the ground between landing and takeoff, should be as short as possible. One important consideration, apart from cleaning the machine, servicing, and fuelling it, is the speed with which passengers take their seats. The time it takes for hundreds of passengers to board the plane can lead to considerable delays and prolong unproductive time on the ground. This is why airlines search for ever-more efficient methods to get the increasing number of passengers that today's aircrafts can hold, into their seats. In the strategies to cut boarding times, the order in which passengers are processed through the cabin plays a crucial role.

According to five Israeli and American mathematicians, the current policy utilized by many airlines is certainly not optimal. It consists in having passengers in the back rows board first and progressively filling the front rows. A typical example would have passengers of rows 25–30 be called up first, followed by rows 20 to 25, gradually moving in five-row increments towards the nose of the aircraft. The idea behind this method is that no passengers in low-numbered rows would block the aisle, thus preventing other passengers from getting to the back rows.

This seems to be plausible; after all, a barrel of beer is also filled from the bottom up. But until recently, no mathematical model existed that could verify the efficiency of this procedure. System engineers turned to makeshift computer simulations, yet these cast doubt on the traditional loading method. Maybe loading an aircraft from back to front is not as efficient as was assumed?

To rely on simulation models was simply not a satisfying method for experts. What was needed was a mathematical model which would allow the explicit calculation of the timespans it takes to load a plane by different methods. A model was eventually found, but it is surprisingly complex. Five mathematicians had to make use of a formula from Einstein's theory of relativity which, until then, had never been used outside of physics. The computer scientist Eitan Bachmat from Ben-Gurion University came across it when analyzing the input and output queues for hard disks in computers.

The model that Bachmat and his colleagues developed considers parameters of the aircraft cabin, the method of boarding, the passengers, and their behaviour. The most important variable, it turns out, is the combination of three parameters: the length of the aisle that is blocked by a standing passenger and his hand luggage—40 cm, for example—multiplied by the number of seats in a row, divided by the distance between two rows. With six seats to a row and a distance between rows of about 80 centimeters, the result would be three. This number means that upon approaching their assigned seats, the passengers of a single row will occupy the aisle space of three rows before they finally sink into their seats.

The problem becomes immediately apparent: if the passengers with row numbers 25 to 30 are asked to board the plane, half of the aisle will be completely blocked and most passengers will be unable to reach their assigned seats until everyone in front of them has sat down. According to the model, the time it takes to fill the cabin grows in proportion to the number of passengers.

The difficulty could be avoided if the policy were to have passengers of row 30, 27, and 24 called up first, followed by passengers in rows 29, 26, and 23, followed by rows 28, 25, and 22, and so on. Passengers boarding the plane in this manner would not block each other in the aisles. To put such a complicated policy in place would be rather difficult, however. (One way to implement it would be to have differently coloured boarding passes, say, red for rows 30, 27, 24, blue for 29, 26, 23, and so on. Then the announcement could simply be "would the passengers with blue boarding passes please go to their seats", etc.)

Surprisingly, the mathematicians' calculations then demonstrated that boarding time could be significantly reduced if airlines would let passengers get onto the aircraft in random order. Just forget the announcements. Even more surprisingly, the model further suggested a still more efficient way to fill the plane: just don't assign any seats at all and let passengers board and pick the seat they want. With this method, or lack of a method, the time required to get people on board and into their seats would only be proportional to the square root of the number of passengers.

But the good news does not stop here. Boarding times could be even further reduced if passengers with window seats boarded first—starting with those in the back rows—then passengers in the middle seat and, finally, passengers in aisle seats. The method might not find favour with families and group-travellers, however, who would be temporarily split up.

A sublimely refined policy would have passengers with window seats situated on one side of the plane board first, followed by passengers who are assigned the window seats on the other side of the plane, with passengers of the middle seats and then of the aisle seats following on.

Chapter 22

Creating Bottlenecks

When flying from Zurich to San Francisco, a traveller usually takes the route through New York City where he transfers to an inland flight to the west coast. But if the North American leg of the journey is fully booked and no additional flights can be added because the route is already congested, the airline may choose to fly its passengers via Chicago. As soon as this option becomes overbooked, passengers can be redirected to their original route. Obviously, it is before the start of the first leg that the decision must be made whether to fly via New York or Chicago. Now, in order to give itself more flexibility, the airline could set up transfer flights between New York and Chicago. These would ease the strain on the network, since there are now more possibilities to channel the travellers. Not only will the strain on the network of air routes be relieved, but costs would be economized and the capacity could be allocated more efficiently.

This sounds both logical and plausible. Unfortunately, it is not always correct. Depending on prices, costs, and capacities, an additional route may make a traffic situation worse. This was first described in the essay "On a paradox of traffic planning" by the German mathematician Dietrich Braess in 1969. The article appeared in the German language and, surprisingly, was not published in English until 2005 when a translation appeared in the journal *Transportation Science*. Since then, the paradox that could arise in street traffic, on the Internet and, generally, in any kind of network, has been named the "Braess paradox" after the article's author. It says that the addition of a new link to the network may reduce rather than improve the network's overall efficiency. For example, the inauguration of a new connecting road in Stuttgart's inner city in 1969 did not contribute to the hoped-for relief of its traffic congestion, but instead, caused outright chaos. Only when the road was torn up again, did relative calm return. In 1990, the closing of 42nd Street in New York City reduced congestion.

The Braess paradox is an example of a situation in game theory, a branch of economics for which Nobel prizes for economic sciences were awarded in 1994 and 2005. In a game, the players who are intent solely on their own advantage have to make choices which, in turn, may depend on the choices of others. In a so-called zero-sum game, one player wins what the other player loses. But the Braess paradox does not relate to a zero-sum game. If the price structure of the network depends in some way on the number of travellers on a particular route, a situation may occur in which all players lose if an additional route is added.

Let us assume there is a traffic jam on a highway while traffic on a parallel street flows steadily. Let us also assume that a street is added now, connecting the blocked highway to the parallel street. What could happen is that just enough drivers decide to join the parallel street to cause a blockage on it, too, while the lower number on the highway was not sufficient to resolve the traffic jam. Now we have two blocked roads instead of one and nobody has gained anything. In fact, more drivers than before have to wait in standing traffic. The paradox can occur if each driver decides for himself whether to take the highway at the outset, and, once at the turnoff, whether to join the parallel street, or to drive on.

Moving to air traffic again, the following, somewhat simplified, situation may present itself. Let us assume that the two routes to San Francisco have capacities of 400 passengers each. That is, a total of 800 people can travel from Zurich to San Francisco. As long as there are only these two routes available, the one via New York City and the other via Chicago, all passengers can reach their destination.

Then, for some reason a route is established between New York City and Chicago. One hundred travellers decide that they prefer the more complicated route Zurich–New York–Chicago–San Francisco, either because prices are better, or because arrival times are more convenient, or because more frequent flyer miles are awarded. Quick off the mark, these 100 travellers make their reservations. Thus, 400 passengers take off from Zurich, one hundred of which switch in New York City to the transfer flight to Chicago. But this leaves only 300 seats for the Zurich–Chicago–San Francisco travellers. Hence, no more than 700 passengers will arrive in San Francisco, 300 via New York and 400 via Chicago. Hence, the introduction of an additional route created a bottle neck. Only when the added transfer flight is removed will 800 travellers again be able to get to San Francisco, 400 via New York City and 400 via Chicago.

Experts believe that the Braess paradox occurs rarely and only under special circumstances. Furthermore, one might have thought that travellers would soon realize that they obtain no advantage by simply changing their route, thereby causing other travellers problems. Therefore, one would hope that should they encounter a similar situation in the future they would not make the same mistake. However, this too is not borne out by facts. Amnon Rapoport, a psychologist from the University of Arizona, known for experimental tests of game theory, analysed whether players learn from their previous mistakes. He provided them with several dozen routes in a laboratory setting, which corresponded to the above example. Without a connecting route, traffic soon settled down to an equilibrium, with each route chosen by around half the players. But, whenever the connecting route was opened, players chose what they thought would be the better route, thus creating the blockage.

Chapter 23

All Flights Lead to
Paris...and Anchorage

One can fly to practically all corners of the earth these days, albeit often by circuitous routes. Stopovers and transfer flights are the bread and butter of busy travelers. The principal question voyagers ask is, how often must they change planes? A study by Louis Amaral, Professor of Engineering at Northwestern University, published in May 2005 in the *Proceedings of the National Academy of Sciences*, analyzed global flight movements during one week of November 2000. Over a half million flights, operated by over 800 airlines worldwide, were studied. With 27,051 direct flights connecting 3,883 cities, only 1.8 tenths of one percent of all city pairs were connected directly. All other connections, more than 15 million, had to be travelled with stopovers.

As the study found, the worldwide net of air connections exhibited relatively few hubs having many connections, and many airports with few connections. In this, the international airline network is similar to other networks, such as the Internet or nets of social connections. It further turned out that the air transportation network is a "small world:" travelers can get from any one city to any other with, on average, 4.4 flights. However, there are also much more complicated flights. The most arduous trip was from Wasu in Papua New

Guinea, to Mount Pleasant on the Falkland Islands. It required no less that 15 stopovers.

The busiest airports, however, are not always the most important ones in terms of the cohesion of the network. With over 100 flight movements per hour, Chicago O'Hare is considered the world's busiest airport. But it offers nonstop flights to only 184 other cities. Meanwhile, the Paris airports (Charles de Gaulle and Orly combined) have direct flights to 250 cities, followed by London with 242, Frankfurt with 237, Amsterdam with 192, and Moscow with 186. At the bottom of the list lie 744 cities which have only a single nonstop flight to another city—Gibraltar for example, or Abu Simbel in Egypt or the Greek island of Mykonos.

But the number of direct connections is not an airport's only measure of importance. Another significant characteristic needed to understand the worldwide air transportation network is the number of shortest trips connecting two cities which run through an airport. Again, Paris leads, with 297 trips between two cities requiring a stopover at Charles de Gaulle or Orly. At second place we find, believe it or not, Anchorage, Alaska! Even though this city way up north has direct flights to only 39 cities, the Ted Stevens Anchorage International Airport serves as a stopover for no less than 279 shortest flights connecting two cities. And the airport at Port Moresby in New Guinea, a place totally unknown to most of the world's travelers, finds itself ranked number seven of the busiest gateways, in front of such busy places as Frankfurt, Tokyo, and Moscow. It serves as stopover for 217 shortest connections. On the other hand, there are 2,491 airports which are not used as a stopover for even a single connecting flight. Thus, if operations at Port Moresby were interrupted, a large part of the international air traffic network would be cut off from the rest of the world. But hardly anyone would notice if the airport at the southern end of the Iberian Peninsula stopped functioning.

The airports, through which many of the shortest connections run, are crucial to the network because they serve as connections to entire regions of the world. From an economic point of view, it is essential that uninterrupted operation at such hubs is guaranteed and not controlled by a single airline or an alliance of companies.

The study is not of interest only to travelers in a rush. Rather, Amaral sees its importance in the detection of an airport's role in the spread of infectious diseases, such as SARS in 2003, later, bird flu, and, more recently, swine flu. The paper shows that the number of flight movements is not necessarily the most relevant variable when studying epidemics. The busiest airports in the U.S., Europe, and Japan do not necessarily serve as springboards for the spread of viruses. Hence, in order to set barriers against the spread of infectious diseases, relatively quiet airports may well play a much more important role than Frankfurt, Chicago, or Toronto. Better keep an eye on Anchorage and Port Moresby.

Chapter 24

Long-Distance Flights are Grounded

In 2007, an article was published in the scientific journal *Nature* which cast serious suspicion on earlier articles, some of which had also been published in *Nature*. However, instead of asking the authors of the previous, erroneous studies to withdraw their former conclusions, the editors allowed them to publish a *sotto voce* rectification, disguised as a further development of the theory and presented as a new conclusion.

The original work examined the foraging behaviour of albatrosses living on Bird Island in the southern Atlantic Ocean. The researchers had attached recording instruments to the birds' feet, which noted when they were wet, which meant that the birds were swimming on the water in search of fish or shellfish, and when they were dry, which meant that the birds were flying in search of food. Once the recording instruments had been retrieved, the lengths of the flights were sorted according to their frequency and statistically analyzed. The results suggested that the albatrosses had, in general, undertaken random short- or medium-length flights, but would occasionally remain airborne for very long periods of time—over 70 hours.

The short flights can be interpreted as a so-called Brownian motion, which occurs in dust particles when they are thrown about randomly in a fluid by the fluid's molecules. Although the jolts forward and backwards and from left to right balance each other out, the average distance between a particle and its original position increases with the square root of time. It was Albert Einstein who made a theoretical prediction of this in 1905, even before he got started on developing the theory of relativity.

What was significant in the case of the albatross study were the few very long time-spans. They suggested to the researchers that after the birds had spent a while foraging, they would embark on a long flight in search of new fishing grounds. Once a new foraging area was found, they would once more search for food in a Brownian motion pattern. The resulting distribution of flights—Brownian distribution interrupted by a few very long journeys—are known as Lévy distributions after the French mathematician Paul Lévy (1886–1971). The distinction between a Lévy distribution and a Brownian motion is important, since, whereas organisms or objects that follow a Brownian motion move forward only according to the square root of time, organisms or pathogens that move according to a Lévy distribution can progress very quickly over long distances.

The albatross study—which until now has been cited 170 times—led to a glut of further studies. They established that jackals, bumble-bees, deer, plankton, apes, and even fishermen behaved according to the pattern described by a Lévy distribution when searching for prey. Quickly, a theoretical explanation was provided. Based on simulations and theoretical deliberations, researchers established that Lévy distributions were the most efficient strategy when searching a large area for scanty food supplies. Thus, the researchers believed, the behaviour of albatrosses, jackals, bumble-bees, and other creatures could be explained in evolutionary terms: animals developed a search strategy whose pattern followed a Lévy distribution, since this offered them the best chances of survival.

There was only one catch: the albatross article was completely and utterly wrong. The studies about bumble bees, deer, seals, and other animals that followed were incorrect too. They all relied on flawed data. Of course, the evolutionary explanations for the phenomena that the theoreticians then offered, did not in any way correspond to reality either, notwithstanding their seeming plausibility.

It was an ecologist working for the British Antarctic Survey, Andrew Edwards, who decided to examine the data more closely. He realised that, in each case, it was the first and the last flight that was measured for each albatross that was unusually long. As soon as these two flights were excluded from the statistical data, the remaining flight times no longer followed a Lévy distribution.

Closer examination revealed that the albatrosses' "flight times" had been measured from the point at which the recording device was connected to the computer. From this moment on, however, it was still a while before the device was actually attached to the legs of the albatrosses. Furthermore, before they eventually flew off, the birds sometimes still spent a fair amount of time in their nests. And after they had returned to their nests, it again took quite a while until the devices were retrieved. All these down-times were measured as flight times by the researchers. But without these two virtual long-distance flights, there was no more statistical evidence of Lévy distribution.

In the studies on deer, there were inconsistencies as well. The researchers had overlooked the fact that the animals spend a certain amount of time grazing any grass they find and digesting the food, before setting off in search of new feeding grounds. These "processing times" were also recorded as travel times. In the case of the bees, the data corresponded to flights from one flower to the next, and not to the search for further flowerbeds. And so on and so forth. Lévy distributions may well be an effective strategy to search for food, but the studies at hand in no way proved that living creatures adopted this strategy in an evolutionary manner.

After all this became known, the authors of the albatross study published a new piece of work, in collaboration with Edwards, in which they admit that the longer flights—indispensable evidence of Lévy patterns—actually did not exist. Instead of hiding away for shame, however, the researchers, in all seriousness, announced their new research results: the remaining data supposedly correspond to a so-called gamma distribution. A retraction of their erroneous earlier article and a meek apology for the fact that countless researchers had been led up the garden path by their sloppy work would have been more appropriate.

Training the brain

Chapter 25

Calculating on the Left Side

Around five percent of the general population is affected by a weakness or outright inability to perform arithmetic calculations. The affliction is commonly referred to as dyscalculia. Afflicted people are largely unable to conceptualize numbers or quantities, have difficulties with measurements and time and spatial reasoning, find it hard, if not impossible, to read timetables or geographic maps, and cannot follow dance steps or count their change. The inability, which is related to dyslexia or dyspraxia, is often diagnosed late in life or not at all. It is quite unrelated to the sufferer's IQ. Indeed, many children and adults who suffer from this weakness, excel in the humanities and in languages. Nevertheless, five- to seven-year-old children who have difficulties recognizing simple number sequences or number patterns, or fail to correctly compare quantities, probably suffer from dyscalculia.

The underlying brain dysfunction responsible for dyscalculia has never been conclusively identified and the cause for the affliction remains largely a mystery. Is it a genetic, inherited, or acquired disability? Is it of neurological origin? A team of seven scientists at University College in London, led by the Israeli neurologist Roi Cohen-Kadosh, sought the region in the brain where dyscalculia originates.

They found it in the right parietal lobe. According to experts in the field, the findings, published in the journal *Current Biology*, may have implications for the diagnosis of the affliction and its management through remedial teaching.

The researchers carried out their study with nine subjects, five of whom were afflicted with dyscalculia; the four other test subjects showed no symptoms. The test subjects were presented with two digits on a computer screen, a "2" and a "4". One of the numerals was always physically larger; that is, it was displayed in a larger font than the other. The subjects had to decide quickly which of the two numerals was "larger." Presented thus, the question was ambiguous, of course, which is why the scientists further specified what they meant. Sometimes they asked for the physically larger digit, at other times they inquired about the numerically larger digit. Then, they measured the time it took the subjects to press a button which gave their answers.

Using magnetic resonance imaging (MRI) to measure the flow of blood through the vessels, the results showed that during the administration of the test, more blood flowed through the so-called parietal lobes—the parts of the brain which participate in the mental manipulation of numbers and sizes. As was expected from earlier studies, nondyscalculic volunteers exhibited faster reaction times when the numerically larger digits were also physically bigger. When the physical size and the numerical value did not correspond, the "normal" test subjects' reaction times were slower. For the dyscalculic subjects, reaction times were slow throughout.

So far, this is not very surprising; one would have expected as much. The really new aspect of the study was what came next. The scientists created a disturbance in the brains of all subjects. Using a special apparatus, they created an electric flow in the parietal lobes of the test subjects for a few tenths of a second, just at the moment when they were to evaluate the numbers. The technique, called transcranial magnetic stimulation (TMS), creates a magnetic field in the brain which, in turn, disturbs the activity of the neurons in the affected region.

The experiment led to a surprising discovery: during TMS-induced disruptions of neuronal activity, nondyscalculic participants displayed the same behaviour as sufferers from dyscalculia, but only if the disturbance took place in the right parietal lobe. When the experiment was repeated through disturbances of the left lobe, nothing extraordinary happened. Obviously, disruptions to the right parietal lobe induced dyscalculia.

Karin Kucian, a neuroscientist at Children's Hospital in Zurich, believes that the British study provides important, but only indirect, information for the diagnosis and therapy of disability in children. With dyscalculia, the entire neuronal network comes into play and, while the right parietal lobe is included, its role is not exclusive. In analogy to the results of the London group, Kucian's research team was able to demonstrate that children suffering from dyscalculia had less grey matter (brain cells) in the right parietal lobe. She also found differences in the anatomy and functioning of other brain regions that are crucial for mathematical activities.

Chapter 26

Losing the Language Instinct

It is widely believed that a necessary requirement for the solving of mathematical problems is the possession of linguistic skills. In the 1920s, Benjamin Whorf, an American linguist and engineer, was the first to propose the thesis that proficiency in a language is a precondition for other cognitive abilities. Dubbed the Whorfian Hypothesis, the thesis has since become a fundamental credo of cognitive science. It is strongly supported by Noam Chomsky from MIT, one of today's foremost linguists. He also believes that it is the command of a language that enables higher cognitive functions, such as doing calculations.

This is, of course, nothing new. For many years neuroscientists have known that the left side of the brain is dominant for language. One would therefore suspect that the left side of the brain also plays an essential role when it comes to solving problems in arithmetic. The assumption has been confirmed by empirical studies which ascertained that it is the left side of the brain that is supplied with blood when solving cognitive problems.

However, in 2005, four scientists working at the University of Sheffield and the Hallamshire Hospital in England submitted evidence that claims the very opposite of the so-called Whorfian hypothesis.

The research team—psychologists, neuroscientists, and communication scientists—showed that people may retain their ability to solve computational problems even after having lost their ability to communicate due to an injury to the left region of the brain. They examined three patients who suffered from "agrammatical aphasia," an inability to comprehend language or to speak in a grammatically correct fashion. The affliction is often caused by a stroke or brain injury.

The three participants in the study, among them a former university professor, were hardly able to communicate with the outside world, neither by speaking nor by writing. They were barely able to utter telegram-style sentences which lacked verbs and prepositions. For example, they could not distinguish between the two sentences "the hunter killed the lion" and "the lion killed the hunter." When it came to constructions such as "the hunter who killed the lion was angry," they failed completely. Surprisingly, however, all three patients had retained their computational faculties despite their severe grammatical impairment. Although the patients displayed difficulties in understanding number words, such as "three" or "twenty-five," they were capable of correctly identifying written digits.

To solve the mathematical problems in the test, the patients had to apply syntactic principles—which they were unable to employ in everyday language—in a computational context. The results were nothing less than astounding. The test subjects were usually able to correctly compute expressions such as $12 - 5$ and $5 - 12$, even though in spoken or written sentences they were unable to distinguish between subject and object. They were also able to solve equations which involved parentheses. Thus, for example, they were able to correctly solve expressions such as $36/(3 \times 2)$, even though they were unable to comprehend clausal sentence constructions, set apart by commas. Expressions containing different sets of brackets, i.e., $3 \times [(9 + 21) \times 2]$, were also calculated correctly. Compare this to the considerable difficulties that even erudite readers often exhibit when trying to comprehend double embedded clause structures in a spoken or written sentence. Thus, the study published in *Proceedings of the National Academy of Sciences* showed for the first time that in the mature cognitive system the possession of grammar is not a prerequisite for the ability to perform mathematical calculations. Apparently, mathematical expressions do not get translated into a language format prior to being comprehended or solved.

The question remains why measurements of blood circulation through the brain show that solving mathematical problems raises the activity in the language region? The scientists suspect that children may require linguistic skills, and the appropriate brain region, in order to *acquire* numerical concepts. This may be the reason why that region is used later on for mathematical reasoning. In conclusion, linguistic mechanisms may merely serve as a memory-hook, a sort of "aide-memoire", in order to become competent in mathematics. Other domains of the brain may be responsible for the specific ability to calculate.

Chapter 27

Information Overload

Many of us find the handling and processing of numerical data difficult. We often feel overwhelmed by the ever-growing flood of information which is so much a part of today's life. That is why hard-to-remember twelve- and fourteen-digit telephone numbers are stored in our mobile phones' memory chips. Numerical data, confusing when presented as a table, is often displayed in graphical form. Financial managers, who must make split-second decisions in real time while simultaneously keeping a watchful eye on stock market figures displayed on up to six screens, have recourse to auditory signals: when particular developments occur on the bourse (European stock exchange), certain melodies alert them to the fact.

But, in spite of such "aides memoires," modern man is burdened by today's onslaught of information: numbers and figures are being rammed home so relentlessly, that many people are unable to make sense of them all. It is important, therefore, to find out exactly what the limits of human information-processing capacities are. How much information and how many variables can the human mind process at any given time? A group of Australian psychologists, Graeme Halford, Rosemarie Baker, Julie McCredden and John Bain published a study in the journal *Psychological Science* in which they showed that even skilled people cannot process more than four variables at a time.

Before embarking on the actual study, the four scientists had to overcome the difficulty of quantifying the information that a human brain can process simultaneously. It so happens that people confronted by complex problems seek strategies to reduce the processing load. A seasoned baker, for example, integrates butter, sugar, and eggs into a single cognitive representation, thereby freeing memory to retain other details.

The scientists, therefore, had to prevent the participants in their study from using mnemonic tricks to condense information. Instead, they asked them to interpret histograms—graphical representations of data—which are often used to represent relationships between different variables. The graphs were designed in such a way that the participants had to comprehend and process all variables before they could attempt to solve certain problems.

Thirty participants were recruited for the study. They were graduate students in psychology and computer science who had experience in interpreting data. Their task was to answer questions about the situations represented in the bar graphs. One of the simplest examples was "People prefer fresh cakes to frozen cakes. Is the preference greater or smaller for chocolate cakes than for carrot cakes?" The problem contains only two variables (fresh vs. frozen, chocolate vs. carrot), so that four bars were required in order to map the information into a graph. The result was as expected. All participants were able to accurately discern the different relative heights of the bars and their responses were correct.

To make things a bit more difficult, the researchers then added another variable—the glazing of the cake. With three variables (fresh vs. frozen, chocolate vs. carrot, glazed vs. unglazed), a bar diagram with eight bars needed interpreting. The results were still good. Close to 95 percent of the questions were answered correctly.

Adding the fat content as a further factor of the cake-making process rendered the problem even more complex. Since the number of bars doubles with each added cake characteristic, participants now had to process the information contained in no less than 16 bars. This proved too difficult for some of the participants. On average, the test persons were only able to solve less than two thirds of the problems correctly.

In the ultimate problem, participants had to interpret five variables, i.e., graphs with 32 bars. It hardly surprises anyone that this time only about half of the problems were solved correctly. This corresponds roughly to the success rate that would have been obtained if the test subjects had thrown out random guesses.

The researchers concluded that the human capacity to interpret quantitative data reaches its limits when four variables are involved in a problem. In general, strategies for reasoning and decision making should, therefore, require the processing of no more than four variables at any one time. More complex tasks should be subdivided into smaller chunks. According to the Australian scientists the special talent of professionals who are experts in their fields apparently consists of their ability to subdivide complex problems into segments, each one of which contains no more than four variables.

Chapter 28

The Case for Mental Arithmetic

Ever since Pythagoras drew his triangles on the sandy grounds of Samos some 2,500 years ago, educators have been searching for the best methods to teach mathematics to their pupils. A case in point was a debate between experts gathered at the 25th International Conference of Mathematicians in Madrid in the summer of 2006. Different approaches used in primary and secondary schools were discussed and disagreement was inevitable. 'Reformers' who take into account social and technical evolution stood against 'traditionalists' who promote paper-and-pencil arithmetic. Lively exchanges ensued, emotions ran high, and not even the most fundamental arithmetic manipulation remained unscathed. Anthony Ralston, for example, an early reformer from State University of New York at Buffalo, vociferously advocated the abolition of paper-and-pencil arithmetic in the classroom. While he admitted that doing computations in one's head is essential for the development of number appreciation, the ability to do mental arithmetic could just as easily be achieved, he maintained, by the use of calculators.

In opposition stood Ehud de Shalit, a number theorist from the Hebrew University of Jerusalem, who is firmly rooted in the traditional ways of teaching mathematics. He believes that teachers need to equip their students from early on with the necessary tools which will then allow them to manipulate mathematical objects such as numbers, shapes, and symbols. As an example he cited long division, carried out with paper and pencil. The technique need not be taught to primary students because it is essential to get on with their daily lives, de Shalit explained. He fully realizes that such calculation can more easily be done by calculator. But it helps the pupils think and conceptualize along mathematical lines. According to de Shalit, long division is, in fact, an absolute treasure trove for teaching, not so much because of its practical value, but because it enhances the understanding of the decimal system and explains how algorithms work. To demonstrate the supposed absurdity of the reformist proposals, de Shalit asked the rhetorical question whether one might not also want to do away with fractions altogether. Fractions can easily be converted into decimal numbers with the help of calculators and could therefore be considered obsolete. But this would be the first step down a slippery slope, he warned his colleagues. Without recourse to a calculator, students would soon no longer know whether $3/7$ or $5/9$ is larger or smaller than $1/2$.

The issue of whether or not to use calculators and computers in classrooms was not the only sticking point pitting reformists against traditionalists. They also had a field day discussing the optimal method to teach students mathematical techniques. Ralston believes that students should be allowed to develop the methods that they feel most comfortable with by themselves. De Shalit quickly dismissed as an illusion that ten-year-old children would be able to discover by themselves mathematical methods which are regarded as some of the greatest achievements by the old Hindi and Arabs. Hence, he wants teachers to concentrate on the tried-and-true standard methods through a drill and practice curriculum. Only after having mastered the standard methods of calculation should pupils be allowed to draw on their own initiative—for example, by exchanging multiplicands.

But de Shalit qualified his strict approach somewhat. Standard techniques are not the most important, and certainly not the only aspect in the teaching of mathematics. To solve actual problems, other abilities are essential: students should be able to distinguish between relevant and irrelevant data, know how to cleverly select the most relevant variables, and be able to translate prose into algebraic formulation. These abilities are indispensable even before the raw techniques are used to solve the problem. In geometry for example, shapes have to be drawn to scale, objects need to be taken apart, and hidden parts must be recognized, before arithmetic can be used for the actual computations.

On one point, both parties agreed: tests are a political matter. They concurred that a fixation on standardized exams hampers teachers in their work. Standardized tests do have their usefulness, traditionalists argue. But it must be stated from the outset what the tests, in fact, are testing. Does it assess acquired knowledge or future potential? Does it measure algorithmic skills or creative thinking? Is the test used for university admission or to analyse different schools or teaching programs?

Reformists, on the other hand, regard standardized exams as an unmitigated disaster. To underline this point, Ralston quotes the "No child left behind" federal act of 2002. The success of the program was measured by standardized test scores. Pressure on the teachers led them to boost the students' ability to do routine manipulations rather than develop their ability to solve problems. Thus, students may have achieved higher test scores, without gaining mathematical ability. Ralston strongly feels that tests should be used solely for diagnostic purposes. There they can be instrumental in determining whether a particular teaching method is successful or not.

Games, gifts, and other diversions

Chapter 29

How Many Moves to Rubik?

Rubik's cube, originally called the Magic Cube, is a three-dimensional mechanical puzzle, invented in 1974 by the Hungarian sculptor and professor of architecture Ernő Rubik. Widely considered to be the world's best-selling toy, it has sold 350 million pieces and its worldwide appeal is such that it retains a cult following, with almost 40,000 entries on YouTube featuring tutorials and video clips of quick solutions. Described by its creator as not so much a toy but "a work of art," it earned a place as a permanent exhibit in New York's Museum of Modern Art and entered the Oxford English Dictionary after just two years.

The classic cube consists of 26 mini-cubes or cubelets on three levels. Each layer of cubelets can be twisted by 90 or 180 degrees. The 270-degree rotation may be ignored since it corresponds, in mathematical terms, to a 90 degree rotation in the opposite direction. Twisting any of the layers independently, the cube can be brought into any of approximately forty-three million trillion possible states of the cube. To solve the puzzle, players have to bring the cube into its original position, with the nine cubelets on each of the six faces showing the same colour.

While even seasoned players are usually happy just to solve the problem, champions of the game want to achieve this goal with the fewest number of rotations. At present it is still not known what the lowest number of turns is that are needed to unscramble the cube from its worst disorder. What we have known for quite some time however, is that at least 17 rotations are required to do so. On the other hand, the London mathematician Morwen Thistlethwaite proved that even the most scrambled cube can be solved in 52 rotations...if you know how.

That no more was known than that the minimum number of moves required to unscramble the most scrambled cube was between 17 and 52, did not let mathematicians rest. Unsurprisingly, therefore, ever since Rubik's cube appeared, mathematicians have tried to outdo each other by improving on the bounds. Eventually the lower bound was brought up to 20 moves while the best upper bound was lowered to 27. However, even this relatively narrow span did not make mathematicians happy. It is only when the lower and upper bound meet that we will know for certain how many rotations are required to unscramble Rubik's cube from its most complex configuration. Only then will mathematicians be able to rest.

In 2007, two computer scientists at Northeastern University, Gene Cooperman and Dan Kunkle, proved that 26 moves suffice to solve any configuration of a Rubik's cube—a new lower-upper bound was reached. Their work, then a true tour de force, celebrated not only a record in the proud history of Rubik's cube, but was a coming-together of many disciplines: combinatorics, algebra, and computer science—the latter both in terms of software and hardware.

Kunkle and Cooperman divided the problem into two parts. In the first step, they considered only those configurations that can be reached by half-turns (i.e., turns of 180 degrees). This subgroup of configurations is quite small by Rubik's standard, containing just 663,552 configurations. The number can be further reduced to only 15,752 configurations after taking into account various symmetries of the cube. The two scientists computed that any of these configurations can be returned to the original state with at most 13 moves.

They then proceeded to analyze the configurations which can be reached by quarter-turns (i.e., turns by 90 degrees), after starting from any of the configurations that can be reached by half-turns. To each of the 663,552 half-turn configurations there correspond 65 trillion additional configurations, reachable by quarter-turns. (65 trillion multiplied by 663,552 equals the 43 million trillion mentioned above.) Running their computer programs on one hundred and twenty processors simultaneously for more than two and a half days, Kunkle and Cooperman came up with the result: 16 moves would return even the most scrambled cube to one of the 663,552 known half-turn states.

So 16 moves plus 13 moves are needed, at most, to arrive from any arbitrary cube configuration to any of the half-turn configurations, and then to the original state. This, however, is two moves more than the previously obtained lowest upper bound of 27 moves. So where is the improvement? Well, Kunkle and Cooperman's program identified only 14,352 configurations which required 29 moves. Since this is a small number for a computer, each of these configurations could be scrutinized separately. In each of the cases, a solution could be found that required not more than 26 moves. Hence, they had a new record.

This was the situation at the end of 2007. Then, developments moved fast. In March 2008, Tomas Rokicki, a Stanford-trained mathematician, posted a paper on the Internet in which he reported that 25 moves suffice. He had also separated the problem into two subproblems. The first sub-problem considered two billion configurations, each of which—that is the second sub-problem—was associated to twenty billion further configurations. He then eliminated a large number of configurations because they were equivalent to other configurations and could, therefore, be ignored. Finally, he ran the problem on a workstation for 1,500 hours.

Just three months later, in June, he struck again. Using the same procedure as before, but this time running the program on the supercomputer at Sony Pictures Imageworks during idle time between productions, Rokicki managed to lower the upper bound to 23 moves. And, using still more brute computing power, he was able to report in August, that the upper bound had now been reduced even further, to

22 moves, a hair's breadth from the highest lower bound of 20 moves. The computations required the equivalent of fifty years of computing time, again donated by Sony Pictures Imageworks. Actually, despite solving more than 25 million billion cube configurations in the course of the computations, none were recorded that required 22 or even 21 moves. But as of this writing an actual proof that 20 moves suffice is still outstanding.

Chapter 30

A Puzzling Puzzle

In the United Kingdom they have become a fixture on the back page of all major newspapers. In America they are popular with Wall Street bankers and desperate housewives, and in Japan they have had a faithful following for the past twenty years. I am, of course, referring to Sudoku, the numbers puzzle. The game claims a global market and appeals to both young and old. All one needs is a pencil. The object is to fill a nine-by-nine grid with the digits one to nine, so that each digit appears exactly once in every column, every row and every three-by-three sub-grid. To make matters difficult, some of the 81 cells of the grid are already filled in. Solving a Sudoku puzzle requires no mathematical acumen, yet it offers food for thought even to seasoned mathematicians.

In Albrecht Dürer's copper engraving "Melencolia," created in 1514, one finds an older and somewhat simpler version of the modern Sudoku puzzle. In the picture's upper left-hand corner there is a so-called Latin square, into which the numbers 1 to 16 are entered. If the onlooker cares to engage in some basic arithmetic, he will note that the numbers in all rows, columns, and diagonals add up to 34.

The magic does not end there, however. The numbers placed into the four corners, into the inner square, into pairs of cells lying on opposite sides, and into many other combinations of four cells also add up to 34. In the face of so much mystical stuff, it comes as no surprise that the numbers placed in the middle two cells of the bottom row are 15 and 14, lo and behold, the year in which Dürer created the engraving. Latin squares, however, date even further back than that. We know of examples going back to Ancient Rome. And in China such magical squares were constructed as long as 5,000 years ago.

Leonhard Euler, the eighteenth century Swiss mathematician, was the first to attempt to comprehend these puzzling squares. The question he asked was, how could 36 officers of six different ranks, from six different regiments be arranged in a six-by-six formation, such that each regiment and each rank is represented exactly once in each row and each column? Yet, try as he might, he could not come up with an answer. Eventually he conjectured that there simply were no solutions for squares with $4n + 2$ columns and rows $(6, 10, 14, \ldots)$.

In 1900, Gaston Tarry, a French civil servant and amateur mathematician confirmed this conjecture for grids with six rows and six columns. Using combinatorial methods, he first reduced the 812,851,200 possible formations to 9,408 squares and then inspected each of them individually. Tarry's systematic method is an example of the method that is used nowadays in computer proofs: in a first step, the set of all theoretically possible solutions is reduced to a much smaller set. In the second step, all members of this smaller set are checked by computers.

In spite of Tarry's breakthrough with the six-by-six grid, the general conjecture as formulated by Euler is false. In 1960, three mathematicians disproved Euler's conjecture for $n = 2, 3$. They showed that one hundred officers of ten different ranks, belonging to ten different regiments could be placed into a ten-by-ten grid and that 196 officers could be arranged in a fourteen-by-fourteen grid.

Producing Sudoku puzzles is not simply a matter of randomly placing numbers picked out of a hat into some of the grid's cells. If too few cells are filled in, more than one solution for the puzzle could exist. If too many cells contain digits, there is no guarantee that a solution exists. How many and which cells need to be filled in, so that there exists exactly one solution, is not yet known. The conjecture is that 16 or 17 of the available 81 little squares need to contain digits.

Overall there are far fewer Sudoku puzzles than there are Latin squares. To understand this, one needs to realize that with Latin squares only the entries in the rows and columns need to add up to the same sum. In the case of Sudoku it is also necessary for the digits 1 to 9 to appear in all three-by-three sub-grids. However, one need not worry that there will soon be a shortage of available puzzles. Initially it was believed that there exist 10^{50} possibilities of combining digits in such a way that they form Sudoku grids. While this is not quite the case, newspapers can continue to rely on a healthy supply of new puzzles with which to entertain their readership. Bertram Felgenhauer, a German IT student at the Department of Computer Science in Dresden, calculated that there are approximately 6.7×10^{21} nine-by-nine Sudoku grids. The number of sixteen-by-sixteen Sudoku grids that exist remains as yet unknown. We do know, however, that Sudoku belongs to the so-called NP-complete problems. This means that as the size of the grid increases, the time required for a computer to find a solution grows exponentially.

Sudokus and Latin squares do not owe their raison d'être merely to the fact that they offer people light entertainment. In fact, many practical situations exist where people rely on the mathematical principles that underlie the puzzle. Sport tournaments are an example. Players need to play each other at various times and in various venues. School time-tables are another case in point. Each teacher needs to get a room for each of his classes during all hours of the school day. Escort services—well, we won't be too precise with this one. In telecommunications, call centers rely on the problem-solving techniques of Sudoku and in computer technology parallel data processing makes use of them. So do social scientists when evaluating questionnaires and agriculturalists, when figuring out crop rotations.

A surprising application of Sudoku and Latin squares arises in medicine. In the same manner as the sums of columns, rows, and diagonals permit the reconstruction of the entire grid, the so-called "Inverse Radon Transform" is used in CAT scans to compute two-dimensional sections of the body.

Chapter 31

Boring Assembly Debates

The initial furor about Sudoku has died down and while commuters still spend their idle time with the puzzle, the real excitement is long gone. But in 2006, an article in the *Proceedings of the Royal Society* reignited interest among the cognoscenti. The article refers to an unlikely source that dates back to the 1770s: Benjamin Franklin. What, one might ask, could this statesman have had to do with Sudoku?

Franklin, one of the Founding Fathers, started his career in public affairs as a clerk to the Pennsylvanian Assembly. As he admitted in his autobiography, the debates which he had to attend proved so incredibly tedious that he spent the time solving number puzzles. Thus, while dreary deliberations droned on, Franklin discovered two very intriguing eight-by-eight squares into which he had arranged the numbers 1 to 64.

As with Sudoku puzzles, the successive numbers are arranged in such a way that all rows and columns add up to the same sum—in Franklin's doodles, to 260. Franklin's squares are not real magic squares, however, because the two diagonals do not sum to this number. But—and here comes some other magic hidden behind Franklin's scribbling—the 32 bent diagonals (four cells in one direction and four in the other) do sum properly to 260. There is more: the numbers in each half-row, in each half-column, and in each two-by-two square add up to 130 (half of 260). The question arises: how many eight-by-eight squares are there in which all rows, columns and bent diagonals sum to 260, and all half-rows, half-columns, and two-by-two squares sum to 130?

In total, there are about 10^{89} possibilities to fill an eight-by-eight grid with the integers 1 to 64. This is an unimaginable number, about a billion times larger than the number of particles in the universe. Franklin's squares, however, need to satisfy several conditions: 32 for the half-rows and columns, 32 for the bent diagonals, and 64 for the two-by-two cells. Number squares that satisfy all of these 128 conditions represent only a vanishingly small fraction of all possibilities. Until recently, only a handful of Franklin squares were known.

It would, actually, be very neat to have a formula for Franklin squares not just of side length eight, but for any side length. Unfortunately, attempts to discover such a formula have remained unsuccessful, even though some special cases were solved by the mathematician Kathleen Ollerenshaw from England. Born in 1912, she served as municipal councillor and later as Lord Mayor of Manchester. (It seems that political debates are particularly well suited for the study of number squares.) To solve her versions of the problem, Ollerenshaw used combinatorial methods. Together with a colleague she was awarded a patent for the use of such squares in cryptography and was appointed Dame Commander of the British Empire by the Queen in 1971.

Since efforts to find a general formula remained elusive, mathematicians had to make do until recently with estimates. Several years ago, for example, Maya Ahmed, a Ph.D. student of mathematics at University of California at Davis, worked out that the maximum possible number of eight-by-eight Franklin squares is less than 228 trillion. As we see, even a vanishingly small fraction can still be an enormously huge number.

Not satisfied by this result, Peter Loly from the University of Manitoba in Canada developed a computer program together with two of his students, Daniel Schindel and Matthew Rempel, to analyze Franklin squares. The article, in which they described what they modestly termed "a very pleasant surprise," appeared in the *Proceedings of the Royal Society* 250 years after Franklin had been accepted to this society as a foreign member.

The professor and his student team decided to use a method called "backtracking," a very efficient search strategy. To apply it, a problem must be organized in a hierarchical structure, like a tree. This structure is then combed for solutions. As soon as one of the conditions—which must hold so that a square can be a Franklin square—is violated, a whole host of potential solutions is discarded. It is as if a large branch with all its twigs had been cut off from the tree.

52	61	4	13	20	29	36	45
14	3	62	51	46	35	30	19
53	60	5	12	21	28	37	44
11	6	59	54	43	38	27	22
55	58	7	10	23	26	39	42
9	8	57	56	41	40	25	24
50	63	2	15	18	31	34	47
16	1	64	49	48	33	32	17

The team fed the problem into a computer and let the program run. The upper bound that had been found by Ahmed was quickly penetrated and the total number of possibilities for Franklin squares shrunk rapidly. After running for a straight fifteen hours, the computer spat out the exact number of Franklin squares of side length eight: 1,105,920. As an added bonus, the program also provided a method for the construction of such squares.

But Franklin would not have been the ingenious man he was, had he not had something else up his sleeve. It must have been an unusually tedious session at the Pennsylvanian Assembly when he constructed a sixteen-by-sixteen square filled with the numbers 1 to 256. All rows, columns and bent diagonals sum to 2056, all two-by-two squares to 514. In Franklin's words, it was "the most magically magical of any magic square ever made by any magician." How many of the 10^{500} sixteen-by-sixteen squares fulfil the requirements of a Franklin square is unknown. According to speculations there should be at least a quadrillion.

Chapter 32

A Step Too Far

Nature is one of the world's most reputable and exclusive scientific journals. Over 90 percent of the submitted articles are rejected. This notwithstanding, a recent canard managed to pass the usually extremely rigorous refereeing process.

In the autumn of 2004, four researchers, three zoologists, and a geographer published a paper in which they predicted how the winning times in the 100-meter race at the Olympic Games would evolve over the next decades. The surprising result: in the Olympic Games of 2156, women will be as fast as men, covering the sprint distance in 8.1 seconds. (To compare: at the Games of 2008, the men's winning time was 9.69 seconds, the women's 10.78.) To add insult to injury, the prediction claimed that in subsequent Games, women contestants would consistently outrun their male counterparts. The fair sex would, by those accounts, turn out to also be the faster one.

The method used by the four scientists for their computations was regression analysis, a technique that determines how numerical data decrease or increase as a variable, say time, changes. According to their analysis, the winning times of men have steadily decreased since 1928 by an average of 0.11 seconds each decade. The corresponding decrease for the women's winning times was 0.17 seconds each decade. The researchers swiftly extrapolated the winning times for the following 152 years and arrived at the incredible 8.1 seconds.

Soon, experts all over the world were seriously speculating about the implications of such disconcerting (for the male competitors, that is) news. Newspapers and journals in many countries tried to fathom the possible physiological reasons for the turnaround in the ultimate competition in track-and-field sports. Could the explanation lie in faster growth of female muscle mass, a smaller supply of testosterone for men or, indeed, the consumption of illegal drugs? Something that never seemed to occur to anyone was to question whether the researchers had used the statistical tools correctly. Apparently, *Nature*'s reputation was too daunting to question anything that appeared in its pages. It would have come close to blasphemy.

But male contestants may breathe a sigh of relief because the authors of the study committed such a gross error in their evaluation of the statistical data that their conclusions are totally invalid. They disregarded one of the basic tenets in regression analysis which says that it is not permissible to extrapolate results beyond the time span used for the actual observations. This, however, is precisely what the authors had done. To realize quite how ludicrous the results were, let us simply extrapolate further. According to the methodology used by the authors, men would run the 100-meter sprint at the speed of light in 2892, while women would then run the distance in *negative* time. Thus, the rules of false starts would have to be adapted, since women runners would routinely shoot past the finishing line before their race had even started.

It doesn't get any better even if you go back in time, either. Achilles, the pre-Olympic hero, would have jogged the 100 meters at a leisurely pace in 43 seconds while Penthesilea would have taken more than a minute to cover the same distance. No wonder Achilles defeated the Amazon in battle. Those were the days when men still were true fellas.

The secret of the erroneous conclusion lies in the fact that trees do not grow into the sky and growth curves have the tiresome tendency to flatten out. In track-and-field sports, this means that winning times for men and women will probably reach a nadir at 9.5 and 10.5 seconds, respectively, for the 100-meter dash. Ignoring this fact leads to statistical "proofs" of a lot of sheer nonsense. Per capita income of the Chinese will soon be higher than that of the Americans, life expectancy will reach 120 years in a few decades time, and the value of a Google share will surpass the United States GNP.

Handling statistics in an improper manner can go beyond the innocent and funny. This was the case in Switzerland in 2004 when the granting of citizenship to foreigners was supposed to be eased. In the run up to the referendum, the issue was widely discussed throughout the country. (In true democratic fashion constitutional change can be effected in Switzerland only with the consent of the people.) An "independent committee against mass naturalization"—in actual fact a xenophobic, if not racist, collection of individuals—published large advertisements in the Swiss press, announcing that by 2040 Moslems would make up 72 percent of the Swiss population if the voting public did not speak out against the alleged rampant naturalization of foreigners. How did the committee arrive at this prognosis?

True, in 1990, around 2.2 percent of the Swiss population was Moslem and this proportion grew to 4.5 percent ten years later. Hence, there is no quibbling about the fact that within a decade the percentage of Moslems doubled. From these two data points, the "independent committee" swiftly concluded that in 2010, nine percent of the Swiss would be Moslem, 18 percent in 2020, 36 percent in 2030 and, horror of horrors, 72 percent in 2040. At this point the committee came to its mathematical senses and put a halt to its prognoses. It does not bear thinking what a continuation of their projections would have implied for the xenophobic among the Swiss voting public: in 2050 the Moslem population would have made up 144 percent of the Swiss population, with the "real" Swiss making up the remaining minus 44 percent. Such a scenario would have certainly taken aback even simple-minded voters.

Chapter 33

Givers and Takers

The young man is in a quandary. What present should he buy his beloved? How can he prove to her that he is serious about the relationship? The rich show-off will try and win her over by opting for an expensive present, say a diamond necklace. The cheapskate will give her just a piece of custom jewellery. The suave dandy chooses the extravagant—an orchid, or tickets for the opening night at the Met. The question, as romantic as it sounds, is actually nothing but a straightforward decision problem that can be attacked with dry mathematics.

This is exactly what Peter Sozou and Robert Seymour from University College in London did. They sought the optimal gift-giving strategy, using game theory and mathematical modelling. Their study was published in 2005 in the *Proceedings of the Royal Society.*

The model Seymour and Sozou built was a courtship game, based on a series of dating decisions. The type of gift the male offers the female is a signal about his quality. The game starts with the man's choice of present. What kind of gift he offers the lady—valuable, extravagant, or cheap—depends on how attractive he finds her. Once a gift is offered, she must decide whether to accept it and grant him an amorous tete-à-tete. Next, the ball is in the man's court again. He must now decide whether to stay with the lady, or whether to call it quits and look for a more suitable partner.

Both parties must be cautious. On the one hand, the gift's value is not immediately apparent to the female. She will only be able to gauge it after she has received it. On the other hand, a diamond ring can easily be turned into hard cash and this would put the man off immediately. Basing themselves on game theory and probability, both partners must adapt themselves to the possible intentions of the partner. The man asks himself whether the lady really likes him or whether she is only keen on the gift. The lady wants to know whether the suitor has a serious relationship in mind or just a brief encounter.

Seymour and Sozou needed to find out which situations correspond to so-called *Nash equilibria*. Such situations were named after the mathematician and Nobel prize winner John Nash, who was made famous even to nonmathematicians in the movie *A Beautiful Mind*. A situation is said to be in equilibrium if neither the man nor the woman could gain anything by unilaterally altering their strategy. Nash equilibria can be computed, though participants in such games do not make any calculations, of course. They find the paths leading toward Nash-equilibria either through the pressures of natural selection, or through the learning process. (For example, when young people adapt to social conventions.) Once players have reached such an equilibrium, there will be no motivation on anyone's part to change strategy. The situation is evolutionarily stable.

The two researchers identified five Nash equilibria. Number five says, for example, the following: "Men offer cheap presents to unattractive women. To attractive women they offer expensive or extravagant gifts, each with a certain probability. Women accept all gifts but only from attractive men. If the present turns out not to be cheap, they decide to mate." The most successful strategy for a man, however, is to offer a potential partner a present which cost him a lot but which cannot be converted into cash. With this the woman is given the twofold message: firstly, she is being wooed by a man who has spending power; secondly, he rates her highly. At the same time the man can rid himself of cynical fortune-hunters since the present does not actually have any real market value.

By the way, extravagant gifts are not a domain exclusive to humans; animals, too, are partial to them. Female peacocks, for example, are quite stricken when a male member of the species exerts much effort to engage in the quite futile, but very stressful exercise of displaying his feathered tail. The biscuit in terms of nastiness, however, is surely taken by the Australian hangingfly (*Bittacus apicalis*). After mating, he tries to steal back his gift—a juicy insect—in order to offer it to another female.

Chapter 34

Who Wins Tic-Tac-Toe?

All over the world children love the game Tic-Tac-Toe, a.k.a. noughts and crosses or hugs and kisses. Played with pencil and paper, or with sticks in the sand, it is a game for two players who are denoted O and X. Players take turns filling the spaces in a three-by-three grid with their marks. The player who succeeds in placing three of his marks in a horizontal or vertical row, or in a diagonal, wins. The game holds only limited interest since most players soon discover that they can force a draw by adopting a strategy that constantly thwarts the opponent—without actually letting them win.

But whoever thinks that the game totally lacks interest is quite mistaken. In 2006, this was demonstrated by József Beck from Rutgers University at the annual Erdős Lecture at the Hebrew University of Jerusalem. Beck, who originally hails from Hungary, presented material from a 600-page manuscript about the game which he had recently completed. With much humor and a Hungarian accent so thick that it did ample credit to the eponym of the lecture series, Beck analyzed questions such as "who wins Tic-Tac-Toe?", "how does he win?" and "how long does it take?".

In the traditional version of Tic-Tac-Toe there are eight different ways of winning: three rows, three columns and two diagonals. In Europe there is a game called "Four In A Row." Here, players place their marks into a four-by-four grid. The winner is whoever manages to put his four marks into a horizontal or vertical row or into a diagonal. Hence, there are ten ways of winning the game: four rows, four columns and two diagonals. When playing the game in three-dimensional space, that is, not on a piece of paper but in a spatial grid, many more ways of winning arise. The mathematically interesting questions now arise, whether the player who makes the first move has an advantage and whether there exists a strategy which would ensure the first mover a win. For spatial Tic-Tac-Toe, mathematicians have been able to prove that on three-by-three-by-three and on four-by-four-by-four grids, winning strategies actually do exist for the first mover.

From the work of Frank Ramsey, a British mathematician who died of jaundice at the age of only 26 in 1930, it follows that higher-dimensional Tic-Tac-Toe cannot end in a draw if the dimension of the game is sufficiently high. In other words: it is impossible to place marks into all cells of the grid without somewhere creating a winning constellation. For this theorem to hold, the dimension must be very high, however. If the side length of the grid is, say, ten, Tic-Tac-Toe would have to be played in a space of about dimension 300 in order to guarantee a win according to Ramsey's theory. At present, mathematicians are investigating whether in this case there is a strategy which guarantees the win to the player with the first move.

Josef Beck obviously likes games. Another one which has kept him busy is the Triangle-Avoidance game. It is played thus: mark six points on a piece of paper. The first player connects two of the points with a red pen, the second player then does the same with a blue pen. The players alternate in drawing their colored lines, all the time avoiding connecting same-colored points to make a triangle. The first player who has no choice but to complete a triangle in his color loses.

Here, too, there can be no ties. According to Ramsey's theorem, one player must win. Once again, the question is whether there exist strategies which inevitably lead to a win. For a set of six points, there are 15 connecting lines. Playing the game to the end, the first player will have drawn a total of eight, the second player a total of seven lines. Hence, unlike in Tic-Tac-Toe, the player who had the first go is at a disadvantage, since he has to draw one connecting line more than the opponent. The question remains whether this opponent can exploit the advantage for a win.

If the game is played with more points, it quickly becomes unwieldy. With 18 points on the sheet of paper there are 153 connecting lines. With so many lines colored red or blue or left uncolored, there are 3^{153} game situations. This corresponds roughly to the number of particles in the universe and the search for a winning strategy is quite impossible. Beck calls this problem, whose treatment is absolutely hopeless even with the brute force of a computer, "computational chaos."

Rightfully, problems such as these should not be handled by computers anyway, but by means of clever mathematical methods. Beck suggested the topic to a Ph.D. student, in the belief that he had handed him a relatively simple topic. After two wasted years, the frustrated student threw in the towel. Unperturbed, Beck went on to suggest the problem to colleagues. Ten years on, the mathematical community is still not much closer to a solution.

Chapter 35

Liars and Half-liars

Most children and adults are familiar with the popular parlour game "Twenty Questions." A player—let's call her Carole—is the respondent and chooses a personality but keeps his or her identity secret. The opponent—let's call him Paul—is the questioner and must find out whom Carole has chosen in her mind by asking her questions which she can answer only with a simple "Yes" or "No." "Is it a man?" "No." "An actress?" "Yes." "American?" "No," and so it continues. If Paul manages to guess the correct person before at most twenty questions, he wins. The most effective way to make guesses is to ask questions that halve the remaining possibilities each time. By asking twenty judicious questions, one can usually identify one person out of a pool of one million people. The reason is that by halving one million consecutively twenty times—and this would be the optimal way of asking questions—only one individual would remain.

There are more difficult variants of the game. One of them would allow Carole to act like an oracle that occasionally lies. (The name "Carole" in this context is meant to indicate the Greek word, oracle.) A mathematician would now ask the following question: how often may Carole tell a lie so that Paul is still able to guess the correct answer by asking a set number of questions? Put differently, how large can the pool of people be so that Paul will still be able to guess the correct individual even though Carole tells a certain number of lies? Of course, Paul either requires more than twenty questions in this case in order to find the sought-after person, or the pool of people from which Paul must choose needs to be smaller.

The problem is named after the Polish-American mathematician Stanislaw Ulam. Joel Spencer from the Courant Institute of Mathematics in New York had been trying to crack it for well over ten years. As it turns out, the solution depends on the precise rules of the game. In one version, Carole is restricted to a maximum proportion of lies at any given point in time. In another version, Carole is permitted to give incorrect answers, but her responses may be distributed differently. Let us assume that twenty questions are allowed and the respondent is permitted to lie at most in a quarter of the cases. In the game's first version, Carole is permitted to lie at most twice within eight questions. In the second version she is permitted to lie five times in response to the first five queries, but—having exhausted her quota of lies—must then stick with the truth in the remaining questions. After lots of fun playing this game, and also quite a bit of mathematical work, Spencer and a collaborator proved that in the game's first version, Paul would be able to determine the correct person only if the number of lies does not exceed half the number of questions. If Carole lies more often, Paul simply cannot win the game. With the rules of the game's second version, Spencer arrived at a different result, however. In this case, Paul already stands no chance of winning if the number of lies exceeds only a third of the queries.

But the game was not over yet for Spencer. Ever the mathematician, he persisted in his research and as is so often the case, his successful search for the solution of one problem bore the seed for another. Together with his doctoral student Ioana Dumitriu, he made the game more difficult. In this version of the game, Carole is allowed to lie, but only if the true response is "no." If the answer is "yes," Carole must truthfully answer in the affirmative. Thus, Carole has become a "half-liar." How large can the pool of people be so that Paul will still be able to determine the correct person by asking twenty questions, even though Carole at times responds with half-lies? We know already that Paul can filter a million people down to the one correct person if Carole never responds with a lie. Spencer and Dumitriu calculated that if one half-lie is allowed, the number is reduced to less than 105,000; if two half-lies are allowed, to 22,000. The threshold drops to 7,000 if Carole is permitted three-half lies.

Having fun with the Ulam problem is only one side of the game, however. A more serious and practical application of the game has proven useful in signal transmission. Computers transmit bits, that is, strings of ones and zeroes. Twenty bits containing ones and zeroes can be said to correspond to a chain of "yes" and "no" responses. If some bits are received incorrectly at the other end of the transmission line due to noise, we are confronted with the Ulam problem. And, if the line transmits the ones correctly but not always the zeros, then we are looking at the half-liar model.

So far, questions and responses alternated in the game. Thus, Paul received feedback before asking the next question and could adjust his queries to the previous answers. Pushing boundaries further, Spence and Dumitriu devised yet another version of the game. Now, Paul has to submit all questions at the start of the game without knowing to which ones Carole would respond with a lie. This means even further restrictions but corresponds more adequately to the situation in telecommunication and computer science. Since zeros and ones are usually transmitted in one direction continuously, without waiting for a response, feedback is limited. The good news is that computer scientists are able to offset this disadvantage with a partial feedback scheme. After the transmission of a certain number of bits, a check digit is sent which permits the detection of errors. Les jeux sont faits.

Chapter 36

Perfect Chequers Ends in a Draw

The game of chequers (the British call it draughts) has been a pleasant way to spend time together for centuries—easy enough to play, yet sufficiently interesting for each party to want to win. During the sixteenth century it was a favourite pastime at the royal court in Spain, though archaeological findings allow us to date it back to the old Egyptians. Now it looks like the fun could be over, however, because computer scientists in Canada have demonstrated—after working on the problem for eighteen years—that the game must end in a draw if two perfectly matched players play against each other.

Chequers is played with 24 pieces on a chessboard. It is a strategy game in which luck plays no role whatsoever. For computer scientists the game serves as a test whether the human brain, relying on its capabilities to grasp situations intuitively and recognize connections, is superior to computers that use brute force to seek winning strategies among billions of possible solutions.

In 1989, Jonathan Schaeffer, a computer scientist at the University of Alberta in Canada, put his money on brute force and began to develop a computer program called Chinook. The program uses search strategies to make the right draw during a game. In the first Man vs. Machine match it was a close call. Playing against Marion

Tinsley, a mathematician and renowned chequers world champion, Chinook lost, albeit narrowly. Two years later, it was a different story. The professor who had lost only seven games during his whole life chose to withdraw from the tournament after six draws. Since then, the jury has been out on whether a machine is superior to man in chequers.

But Schaeffer continued to develop his program further. It was a daunting task, since there are 500 billion billion (5×10^{20}) possible situations that could arise in the course of a game. The normal brute force approach—search for the optimal move among all the game settings—is quite infeasible.

So it was left to man, in other words, to Schaeffer and his colleagues, to develop clever algorithms. First, the team constructed a database in order to analyze the endgames with ten pieces, or fewer, left on the board. This reduced the number of situations that had to be analyzed to a mere 39 billion endgames. They were to be classified according to whether they ended in a win for black, a win for white, or in a draw. It took seven years—from 1989 to 1996—to classify all endgames with eight pieces or fewer. At that point, the team put a temporary halt to their work and waited until a faster and more powerful computer appeared on the market. When it did, the team was able to analyze the nine-piece and ten-piece scenarios. At times there were up to 200 desktop computers in use, whirring away at the problem.

In the next phase, the scientists analyzed the situations that could arise three moves after the opening. A specially developed algorithm sought the moves that offered both players the best chances of winning. Lo and behold, Schaeffer found that if both players played a perfect game, they would invariably end in a draw.

So far, chequers is the most complex game that has been analyzed by computers. Will chess be the next game to bow to the powers of the computer? For the foreseeable future, say experts, there is no danger of this happening. As complex as chequers is, chess is much more complex still. The number of possible positions in the Royal Game is roughly 10^{40}.

For this reason, Schaeffer has meanwhile turned to poker. The challenges in that game are special in that, unlike with chequers or chess, the players do not possess all the information, quite aside from the bluffing factor. In a game between Schaeffer's computer programme "Polaris" and two of the world's best professional poker players, the humans still won. But it was a very narrow win—Humans: 2, Polaris: 1, and one draw.

Choosing and dividing

Chapter 37

The Talmud—A Precursor to Game Theory?

When you see an orthodox Jew in Jerusalem with a white beard and a skull-cap on his head, you would assume that he was a rabbi. You may be taken aback when you find out that the man is the 2005 winner of the Nobel Prize in Economic Sciences, Robert Aumann. Born in Germany in 1930, Aumann's family emigrated to America at the outbreak of World War II, when he was eight years old. As an adult, he moved to Israel which he now calls home. Aumann is a deeply religious Jew, fully committed to the Jewish State. The fact that he chose to align himself with the Israeli right wing makes him somewhat of an outsider among his more liberal-thinking colleagues.

Aumann's research deals with game theory which allows the analysis of economic behaviour in a scientific, mathematically rigorous manner. Ever since becoming Professor Emeritus at the Hebrew University, Aumann has been a member of the Center for Rationality in Jerusalem.

The professor is known for his personal warmth and good sense of humour. Around town, in Jerusalem, everyone greets him by his Hebrew first name, Yisrael. The author of this book met him personally for the first time when he arrived in Jerusalem as a student. Without further ado, Aumann invited him to his house for a talk. Already famous then, Aumann was well aware that the student opposite him was in awe, but immediately sought to make his guest feel comfortable. He asked his young son to bring a cup of cocoa and with this, the ice was broken. Tragically, this son would be killed as an Israeli soldier a few years later, in the 1982 Lebanon War.

Three years after this loss, Aumann and his colleague Michael Maschler published a treatise in the *Journal of Economic Theory* on a Talmudic question that had been answered by the sages in a way that hitherto nobody had understood. The two mathematicians discussed an age-old tractate, in which a husband's inheritance had to be divided among his three surviving wives. Aumann dedicated this piece of research to the memory of his son.

The problem is of a man who had decreed in his will that his three wives should receive 300, 200, and 100 Zuzim, respectively. After his death, however, it turned out that the man's entire estate consisted of only 200 Zuzim. How was the inheritance to be divided? Today's executors of a will would allocate to each woman a pro-rated share, half of the inheritance to the first wife, one third to the second and one sixth to the third. The Talmud, however, came to a different solution: The first two wives were to receive 75 Zuzim each, and the third wife 50. How did the Jewish sages arrive at these mysterious numbers?

The problem, according to Aumann and Maschler, would be best understood in the light of modern game theory. The division decreed by the Talmud corresponds to the so-called nucleolus of a game. Let us say that two creditors are owed $300 and $200, respectively, in a bankruptcy case but that the entire assets of the bankrupt company are worth only $350. The parties could argue in a Talmudic court as follows: the first creditor, Leo, claims his undisputed $150 from the $350 since the other creditor, Linda, would in the best of cases not receive more than $200. Using the same argument, Linda could claim an undisputed $50, since Leo would, at best, not receive more than $300. Once the sums of $150 and $50 have been distributed, there remain $150. This sum would be divide by the Talmudic judges among the two creditors in equal parts. Hence, Leo would receive $225, Linda $125. A proportional distribution, as practiced by modern judges, would have awarded Leo and Linda $210 and $140, respectively (60 and 40 percent of the remaining assets). The distribution of $225 and $125 is known in game theory as the problem's nucleolus.

With three or more creditors, things become trickier. But Auman and Maschler developed a method by which the nucleolus can be found in such cases, too. By definition, the solution has to fulfill the following procedure: calculate the sum of the shares that you believe any two of the creditors would receive by the Talmudic method. Then check to see, if the Talmudic division does, in fact, add to that sum.

To illustrate, take the two widows who claim 200 and 100 Zuzim. According to the Talmud, they should receive 75 and 50 Zuzim, that is, a sum total of 125 Zuzim. Now, let us check if these numbers fulfill the criterion: the first widow can claim an undisputed amount of 25 Zuzim (since widow number 2 would, in the best of cases, only receive 100 from the estate of 125 Zuzim). Widow number 2, on the other hand cannot lay claim to anything but zero (since widow number 1's claim of 200 Zuzim exceeds the total amount available). After the 25 Zuzim have been awarded to widow number 1, the remaining 100 Zuzim are split equally between both women. Hence, the shares in the estate, as recommended by the Talmud, correspond to the nucleolus.

Aumann likes to point out that the Talmud is a treasure trove for economic theory. Fundamental concepts like risk aversion, the invisible hand, free competition, and standardization of weights and measures, can all be found in the Jewish code of law, written nearly two thousand years ago.

Chapter 38

How the Cake Crumbles

This is how the Bible tells the story of how Abraham and Lot divided the Holy Land among themselves: near the town of Beth-El the uncle drew an imaginary line leading from the North to the South. "If you go right, I will go left," Abraham told his nephew, "if you go left, I will go right." The two parts among which his nephew now had to choose were not of equal value. The eastern part, the Jordan valley, comprised rich, well-watered lands and was covered in lush vegetation. The western part consisted of unknown highlands. Apparently, Abraham's concern was not to divvy up the land fairly into two parts of the same size and value. For him it was sufficient that the lands eastward and westward of Beth-El were, in his eyes at least, equally valuable. So he left the choice to his nephew.

Lot, concerned about his immediate material gain, predictably chose the East. He was happy since, strictly speaking, he had gained more than his fair share. For his part, Abraham, having secured Canaan, also felt content since he had been indifferent between the two parts of the land anyway.

So, according to their own assessments, both Abraham and Lot may consider themselves lucky. Each of them walked away with at least half of the total value. What the Old Testament so beautifully illustrates is a procedure that economists and mathematicians call a fair division. Neither of the two men felt that he was put at a disadvantage. Actually, this method may be familiar to us from our childhood days. When Peter and Tom divide a piece of chocolate, Peter may split it and Tom chooses. Or Tom splits and Peter chooses. In both cases the chooser can be sure that the splitter will do his best to divide the object in a fair manner. That is why, after getting their respective pieces, neither will envy the other one. But what if the bar contains a hazelnut? What if the object that must be divided is—as in the case of the Holy Land—not homogeneous? Often, the counterparties can compensate each other through monetary payments if the estimate that the components they are to receive are of unequal value. But how does one determine the value of a hazelnut? Or take a divorce. Assets can be sold and divided. But what is the value of visitation rights with the children? And what is to be done with a chocolate bar that must be divided among three or more kids?

The question turns out to be considerably more complex than it appears at first glance. During the 1940s, the Polish mathematician Hugo Steinhaus (1887–1972), worked on the problem. He demanded that a fair division, first of all, be proportional (each party must be convinced that he has received at least the fraction he is due) and, second, not give rise to envy by one of the parties (neither party must prefer the other one's share to his own). Steinhaus then proved that fair procedures to divvy up an object among any number of parties must exist. Unfortunately this is all he managed to do. He was unable to devise an actual procedure which would perform the fair division, except for the case of three participants. And even it was not envy-free.

Only in 1962 did John Selfridge of Northern Illinois University and John Horton Conway, then at Cambridge University in England, independently find a practicable method for the three-partner case. It was both proportional and envy-free. However, it was much more complicated than if only two parties were involved. Let us assume that Albert, Beth, and Charlie want to divide a cake. Albert has the first go and divides the cake into what he believes to be three equal parts. Beth, with a keen eye, does not believe that Albert did a fair job and starts cutting away a sliver from the piece she thinks to be the biggest of the three. She measures carefully while cutting to make sure that this piece is now equal in size to the second biggest. Both of these two pieces would be acceptable to her. Now Charlie chooses the piece from among the three which he likes best. He is happy since he got what was, in his eyes, the best piece. Beth then chooses either the piece which she trimmed or the formerly second largest. She will feel no envy since, after all, she gets one of the two portions that were acceptable to her at the beginning. Albert is the last to choose. Since he was the one to slice the cake in the first place into what he thought were three equal parts, he has nothing to complain about either. So, all three are content, but what happens with the sliver that Beth had cut off? Well, now a new round of slicing, trimming, and choosing starts, just as before, but this time using the sliver that has been left over from the first round. In principle, this procedure never ends, with smaller slices, slivers, and finally, crumbs being divided, ad infinitum.

After the end of World War II, at a time when Steinhaus was still struggling with the problem, there was a famous case where Selfridge's and Conway's method was already used, albeit unwittingly. In February 1945, Great Britain, France, the Soviet Union, and the United States decided in the city of Yalta, in Crimea, to eliminate Germany as a superpower. They would divide the country, with each of the allies receiving one piece. But they found it difficult to reach a satisfying, envy-free solution. Only after removing Berlin from the Russian sector and treating it like the surplus cake slice—by dividing it, in turn, into four pieces—could they reach agreement. (Maybe, just maybe, the Berlin example could serve as a blueprint for the two-state solution in the Middle East, by removing the Holy City from the disputed real estate.)

Strictly speaking, the procedure of dividing the trimmings, and the trimmings of the trimmings, etc., is only valid for three participants. In 1995, however, two mathematicians, Steven Brams and Alan Taylor, discovered an algorithm that allowed the extension of the method to more than three players. Unfortunately, a large number of cake cuts and trimmings are then required, with the number doubling with each additional player. Nevertheless, the professors hurried to have their fair-division procedure patented. Anybody keen on finding out how this method would have impacted on the division

of assets in the divorce of Ivana and Donald Trump, can take a look
at patent number 5,983,205 on the website of the U.S. Patent Office.

Chapter 39

Spoilt for Choice

The axioms of set theory, formulated at the beginning of the twentieth century by Ernst Zermelo, Abraham Fraenkel, and Thoralf Skolem represent the foundations of modern mathematics. However, one of the proposed axioms, the so-called axiom of choice, provoked controversy. On the one hand, some mathematical theorems can be proved only by having recourse to it. On the other hand, many purists could not reconcile themselves to an axiom that postulates a function that chooses certain elements from among a set of elements, but fails to specify how the choice is to be made.

Nowadays, the vast majority of mathematicians and scientists make liberal use of this axiom in their everyday work—usually without even being aware of the fact. But some theoretical mathematicians prefer to make do without the axiom of choice. They claim that it permits the proof of seemingly absurd results. For example, the axiom of choice has been used to prove that a solid sphere can be chopped into parts, which can then be reconstructed as two new spheres, both of them identical in size to the original sphere. Which school of thought mathematicians choose to follow can have considerable consequences. Saharon Shelah from the Hebrew University of Jerusalem and Alexander Soifer from the University of Chicago proved that even the answer to concrete mathematical problems may depend on whether one accepts the axiom or not. Their work showed that a world with the axiom of choice differs more from a world without it than was heretofore believed.

Imagine a family of sets, each of which contains a number of objects. The axiom of choice says that from each of these sets one object can be picked out. Every morning we make choice-decisions when we get dressed: we pick out one shirt from a set of shirts in the cupboard, similarly one pair of trousers from a set of trousers, one sweater, etc. Since the cupboard holds only a limited number of clothing in each category, our choices can be made explicit. For example: "pick the blue shirt on the upper right shelf."

When it comes to infinitely large sets, one runs into problems. Actually, this need not always be the case. As the philosopher Bertrand Russell pointed out, one can always pick one shoe from each of an infinite number of pairs of shoes. The choice rule could simply be "pick the left shoe from each pair." But when confronted with an infinite number of pairs of socks, there is no explicit way to make a pick. In order to select one sock from each of an infinite number of pairs, the axiom of choice must be postulated. Another example is that in school with an infinite number of classes the best pupil from each class can be picked. But faced with an infinite number of matchboxes there is no rule to choose a specific match from each of the boxes. To recap: for pairs of shoes or school classes the axiom is not required, because specific choice rules can be formulated. But for socks or matches, where there is no way of identifying particular elements from the sets, one must have recourse to the axiom of choice. Only then is it permissible to say "choose one sock" or "choose one match."

In the 1960s Robert Solovay showed that the axioms of Zermelo, Fraenkel, and Skolem, in conjunction with the axiom of choice, implied the existence of nonmeasureable sets. This means that an alternative system of axioms is conceivable. Formulated by the Swiss mathematician Paul Bernays, it requires only a weaker version of the axiom of choice but, in exchange, postulates the so-called Lebesgue measurability of sets. (The Lebesgue measure gives the "size" of sets, comparable to lengths, areas, or volumes in Euclidean space.) Both systems of axioms can be used to derive mathematical theorems, but they are mutually exclusive: only one or the other system of axioms can be valid.

In a series of papers Shelah and Soifer show that it is not only of philosophical interest whether or not the axiom of choice is accepted. It may also have an impact on the results of concrete problems. Thus, the axiom of choice assumes significance similar to that of the parallel axiom in geometry. Since the time of Euclid, mathematicians had been convinced that the parallel axiom just had to be valid. Without it, many theorems in geometry, confirmed by everyday experience, would have been unprovable. But in the nineteenth century, Bolyai, Lobatschevsky, and Gauss came along and showed that there were geometries that made do without the parallel axiom. With this they opened up new worlds. By dropping the parallel axiom they showed that alongside Euclidian geometry with its plane geometry, other geometries exist that are valid in curved space. (Einstein, for example, required "non-Euclidean geometry" to develop the Theory of Relativity). In the same manner, Shelah and Soifer showed that there are several realities, depending on whether the axiom of choice is considered valid or not.

The two mathematicians started out with a problem that had been formulated in 1950 by then-just-18-year-old student Edward Nelson (today a professor at Princeton University). Nelson considered all real points in the plane and asked how many colours it would take to paint each of them in such a manner, that no two points at a certain distance from each other, say one centimetre, are of the same colour. It can be shown relatively easily—without having recourse to the axiom of choice—that at least four colours are required and that seven suffice. But do we need four, five, six, or seven colours? In three-dimensional space the same problem exists. There it is only known that the required number of colours lies between six and fifteen.

Shelah and Soifer started out with a somewhat simpler problem, namely the colouring of points along the real line. They demanded that two points be of different colours if they are at a particular distance from each other. (For reasons we shall not go into, they set that distance as a multiple of the irrational number $\sqrt{2}$.)

To give a taste of how to calculate the number of colours required under the assumption that the axiom of choice is valid, imagine an infinitely large class of students, all standing in a row. According to the axiom of choice, there is one student who is at the top of his class. The other axioms then allow us to determine at which distance each student stands from the class leader. Depending on whether, for example, the distance is an even or uneven number, the student receives a particular colour. Therefore two colours are required to colour the students in that manner.

Let us now drop the axiom of choice and, in its stead, postulate Lebesgue measurability. To demonstrate the idea of Shelah and Soifer's proof let us consider a box of finite size filled to the brim with infinitely small, identical matches. Since the matches are indistinguishable, the above method cannot be applied. But let us assume for now that the matches can be painted with n colours. Now, allocate each match according to its colour to one of n classes. Based on the Lebesgue measure, the sizes of the classes can be determined. Shelah and Soifer demonstrated that since the matches are infinitely small, the classes are always of zero size. And since the combination of zero-size classes also has zero size, the collection of all matches would have size zero. But this contradicts the fact that the box has finite size. Hence, the assumption must be wrong: n colours do not suffice to colour all matches, no matter how large n is.

In their subsequent work, Shelah and Soifer provide examples of colouring problems for the plane and for three-dimensional space. They lead to the same problem: positing the axiom of choice implies one result; dropping the axiom and positing Lebesgue measurability in its stead implies another. Nelson's question remains unanswered and Shelah and Soifer showed why this might be the case. The alarming conclusion is that many problems simply do not have one single solution. So, even though the axiom of choice might be accepted by the vast majority of mathematicians, Shelah and Soifer brought its inherent complexities to the public's attention. This, in turn, reinforces the requirement to own up to one of the two systems of axioms. One may not simply assume the validity of the axiom of choice. Thus, while spoilt for choice, we are left with no choice but to choose.

Chapter 40

Selecting the Best Pope and the Best Song

It must have been a tense feeling back in April 2005 when 115 cardinals withdrew to the Sistine Chapel in Rome to elect the new Pope: the holy men had no idea how many ballots they would have to sit through, since to win the election a cardinal required a majority of at least two thirds of the votes. Given internal rivalries, squabbles, fierce competition, and antagonism, the election process could take days.

An election of just slightly less significance occurred a few weeks later, when the winner of the Eurovision Song Contest was to be determined in Kiev, in May of the same year. Twenty-four singers had gathered in the *Palats Sportu*, anxiously awaiting the result. Whose song would be selected as the best (or at least the least offensive) one? Here too, a cleverly devised system was going to determine the outcome.

One of democracy's greatest achievements is the principle that every citizen has the right to a vote in an election. But there is a definite drawback to this widely practiced method. Since each citizen can cast his or her vote only for the preferred candidate, it is not clear as to how much he or she actually prefers the particular candidate. Some voters may prefer a contender only slightly to the competitors, while for others this same candidate might be miles and miles above the rest. In short, the principle of "one man, one vote" (and of course, "one woman, one vote") cannot reflect the candidates' true ranking. This leaves the very real possibility that the majority of voters will settle for a compromise candidate whom nobody really wanted.

In 1770, the naval officer and mathematician Jean-Charles de Borda (1733-1799) suggested an election method to the *Académie des Sciences* which would allow voters to better express their preferences. If, say, five candidates run for election, each elector would award four points to his or her preferred candidate, three points to the second choice, two points to the next candidate, one point to the next, and finally, zero to the last one. The candidate who accumulates the most points is elected. The system was dubbed the Borda-count.

Not everyone approved of the Borda count. Among the vocal critics was the mathematician and politician Marquis de Condorcet (1743–1794). He felt that Borda's method invited intrigues. Since Condorcet inhabited the worlds both of science and of politics, he immediately understood that once a group of electors becomes aware that a competitor to their favorite candidate stands a good chance of winning, they would form a coalition to stop his aspirations. By awarding him zero points in concert, they could ruin his chances of winning. A possible result of such an intrigue could be the election of a third-ranking candidate, who never was a real option and would merely win on the basis of compromise alone.

Thus, as an alternative to the Borda-count, Condorcet put forward his own suggestion. In a series of showdowns, every candidate would be pitted against every other candidate. In each such duel the winner is chosen by majority vote. Obviously, a candidate who beats all competitors must be the winner. But even though it would identify the most worthy candidate, Condorcet's system has its own drawback. For one, it would be very time consuming. With 115 cardinals it could take up to 6,555 showdowns to choose an uncontested pope. (Even if the process of showdowns would, by coincidence, start with the superior candidate, who beats all competitors, it would still take 114 rounds to declare him the winner.) At the Eurovision Song Contest this would involve listening to the inane tunes repeatedly, ad nauseam. But an even more serious argument against Codorcet's method is that most often no candidate exists who would beat all the competitors. Generally, even the most outstanding candidate loses at least some of the showdowns.

In short, there is no ideal voting method. While Borda's system guarantees a winner, it opens the doors to manipulations. And while Condorcet's system is immune to manipulations, there may be no Condorcet-winner.

In chess tournaments the Condorcet method is standard, with all contestants playing against all others. In tennis matches the competitors are arranged in a tree-like formation, with the number of successful competitors being halved in each round. European soccer championships employ a two-stage process to find the winning team. For the qualifying round, groups of teams are selected by lots. Condorcet's method is then used within the groups, with each team playing against each of the others. The teams in each group with the most wins are then invited to the finals where they compete as in tennis tournaments, with half of the teams falling by the wayside in each round. To judge the songs in the Eurovision Contest, a version of Borda's method is used: each country's jury awards no points to the contestants with the most horrible songs. Next, they allocate between one and ten points to the ten songs they like better. Finally, they give twelve points (the notorious *"douze points"*) to the song they deem to be the best one.

Why twelve? The answer is mundane: the number is picked out of a hat. The Eurovision winner could quite easily be a different one if, say, the juries were allowed to allocate eleven or thirteen or twenty points to their favorite song. On the other hand, there is a reason for the stipulation that a two-thirds majority be obtained by the winning pope: at least half of his supporters would have to go over to the opposition candidate, if there is one, in order for him to achieve the required majority. No quibbling here, neither from a mathematical nor from a religious point of view: the Pope is, after all, infallible.

Money, and making it

Chapter 41

Follow the Money

On the day after New Year's Eve in 2002, many European countries woke up to a new currency, the euro. While this meant feverish calculations for the ordinary shopper, for economists, financial experts and statisticians a unique occasion presented itself to do some research. The faces of the euro coins which show the denomination are identical in every country. But the reverse sides boast separate, nationally inspired images. A veritable kaleidoscope of European art, history, and music presents itself: paying for a paella in Spain may leave you with change showing the Austrian Mozart; a Parisian café owner may find coins with Leonardo da Vinci's *L'Uomo* in his till; a Dutch waiter may receive a tip that carries the French motto, *Liberté, Egalité, Fraternité.*

The amount of coins minted in each country corresponds to its economic significance within the European market. Out of a total of 65 billion euro-coins, 32.9 percent originate in Germany, France, Italy, and Spain; each have a share of some 15%, while Luxembourg can claim only 0.2 percent. Approximately 100 coins were issued per citizen on 2 January 2002. Because of the differing coins, the change-over in currency allows researchers to track the paths of coins across borders and to study the currencies' circulation.

A study carried out by the Dutch National bank found that any person carries on him an average of 15 coins. This leaves the other

85 coins in the tills of banks and shops. Calculations which take the travelling habits of Europeans into account show that some ten percent of the coins are exported from each country each year, while about the same amount is brought back home. With time, the coins get mixed together in the European countries until an equilibrium has been achieved. Then the coins will be distributed among all countries in proportions that correspond to their individual coinage. One of the questions scientists try to tackle is how long it will take for this to happen. And this is where opinions differ.

Dietrich Stoyan, a professor of statistics in Germany, developed a mathematical model which consisted of nearly 150 differential equations. As variables he used the mobility of travellers from the different European countries, the behaviour of professional commuters, the activities of coin collectors, the preferences for vacation destinations, the family ties of cross-border commuters, and so on. The model also takes into consideration that there is more money migration in the summer months due to holiday travelling, and that following the ski season, Austrian euros accumulate in the flat countries. Most of the other models that have been developed differ only in terminology.

To verify the validity of their models, scientists rely mostly on reports supplied by volunteers who report from time to time the number and provenance of coins they find in their purses. Of course, this method is statistically not very reliable since observers tend to file their reports mainly when they come across an especially high number of foreign coins in their purses. There was another factor which threw a wrench into the research and which was underestimated at first: coins making their way into foreign countries tended to disappear from circulation at first, because collectors preferred to hoard rather than use them. Nevertheless, the data that was collected suggested a definite move from the north to the south. The finding can be explained by the fact that more Germans prefer to go to Spain than Spaniards to Finland.

The point is to establish which of the various models best describes the coin-mixing process. Since the models give different predictions about the exact time when a complete mix of euros has been obtained, the speed at which the mix develops can serve as a test for the research. Stoyan's model suggests that a complete equilibrium should be achieved by 2020.

The study of the migration of the euro is not only of interest to economists, however, but has useful applications in other areas as well. For example, the results can shed light on how epidemics spread, how rumours propagate, and how plants invade foreign habitats. Should Switzerland or Great Britain enter the currency union one day, a new study could be carried out. It would investigate how an organism reacts when a fresh source of infection arises or what happens when the membrane of a cell breaks and allows microorganisms to enter.

Chapter 42

Earthquakes, Epileptic Fits, and the Stock Market Crash

The vagaries of the stock market are governed by the supply and demand of shares by rationally thinking investors—at least that is the theory propagated by classical financial market economists. But a relatively new breed of scientists, the so-called econo-physicists, beg to differ. They see market participants as a community of autonomous agents who interact in ways similar to gas molecules.

The increased frequency with which global crises grip financial market puzzles both economists and financial market theorists. According to their models, exceptional events such as the crash of 1987, the bursting of the dot-com bubble in 2000, the sharp rise in oil prices in 2007, or the most recent crunch, should happen much less frequently than they do. The models that have been developed so far have apparently not captured the reality. It is small wonder, therefore, that scientists outside the financial and economic arena felt called upon to take a closer look at economic science. The most recent group of scientists to do so are physicists who, under the banner of "econo-physics," attempt to describe markets with methods borrowed from statistical mechanics.

But they were not the first. Earlier, psychologists and behavioural scientists had tried to understand the economy's mysterious ways. Based on surveys and controlled laboratory experiments, these scientists tried to find out how people make financial decisions. Daniel Kahneman and Vernon Smith were among the pioneers in this area and were awarded the Nobel Prize in Economic Sciences in 2002 for their work. More recently, neuroscientists appeared on the scene. Equipped with machinery that measures the blood flow in the brain's blood vessels (using a technique called magnetic resonance imaging, MRI), they determine which regions of the brain and which emotions are at work when a person engages in an act of buying and selling. And then came the physicists. In contrast to their colleagues in other fields, they do not study single persons and their actions, but, rather, view market participants as parts of a whole, as "agents" who interact, similar to molecules in a gas. In an attempt to understand the behaviour of markets, econo-physicists employ methods borrowed from statistical mechanics which had originally been developed to trace macroscopic attributes of a gas, such as pressure and temperature, back to the microscopic behaviour of molecules.

According to the traditional theory of financial markets, prices on the stock exchange are determined by the market participants' supply and demand of shares. Investors, intent on maximizing their wealth, move towards the shares' "true" value by making rational decisions, based on the current state, and on the future expectations, of fundamental economic variables. In order to find the mathematical solution to their decision problem, "rational" investors use the classical methods of differential calculus. Jean-Philippe Bouchaud, professor of physics at the Ecole Polytechnique in Paris and head of research at Capital Fund Management, believes that this point of view is in need of an overhaul. Financial engineers, he claims, have put too much faith in untested axioms and faulty models. The customary theory relies, erroneously he claims, on concepts such as the invisible hand, rational investors, and efficient markets. Over the decades, these concepts solidified into axioms from which economists are reluctant to part, in spite of empirical experience indicating that they are not valid. Such ossified thinking can be dangerous. Bouchaud believes, for example, that the deification of the free market has led to deregulation and, tragically, to the most recent crisis.

Yet another example of obsolete economic models are distribution problems. In classical economic theory the representative agent who optimizes his wealth is a core concept. But if all consumers were representative agents, explains Thomas Lux, Professor of Monetary Economics and International Finance at the University of Kiel in Germany, no gaps in income and wealth could develop. Other experts argue that the perfect market—according to which the price of a share reflects all available information—only exists in theory.

And yet, old habits die hard and economists cling to the traditional concepts. This is where the time-honored science of physics may be able to teach the relatively young science of economics a thing or two, claims physicist Bouchaud. If nothing else, he says, physics has centuries of history behind itself and has learnt to deal with setbacks and failures. Physicists have thus learnt that they must, on occasion, eat humble pie and discard useless theories. This approach to theory building, a sine qua non in scientific research, is apparently still foreign to many economists.

The happy union, call it symbiosis, between physics and financial theory began in 1900 when the French mathematician Louis Bachelier presented his Ph.D. thesis to the University of Paris. He investigated movements on the stock exchange with the help of methods which he had developed precisely for this purpose—mind you, five years before Einstein independently developed the exact same methods—in order to describe the erratic movements of particles in a fluid, the so-called Brownian motion.

But a serious problem would henceforth trouble future observers of financial markets. One of Bachelier's basic assumptions is faulty. He postulated that price changes follow the so-called Gaussian normal distribution: most changes in stock prices are small and plot nicely on a bell-shaped curve. This assumption is wrong, however, in a crucial point. While small and medium price changes generally do follow this pattern, extreme events, such as stock market crashes or sudden price increases, occur much more frequently than one would expect according to the bell curve. Statisticians say that compared to the

Gaussian bell curve, the actual distribution of price fluctuations has "fat tails."

A group of physicists, led by H. Eugene Stanley from Boston University, sought other distributions which would be better suited to describe the behaviour of stock markets. Eventually they came across the so-called scaling laws and power law distributions which Benoît Mandelbrot, the founder of chaos theory, had used to measure coastal lines, the surfaces of cauliflowers, and, also, price fluctuations on the markets for raw materials. Econo-physicists point out that many natural phenomena are based on such distributions. Didier Sornette, for example, Professor on the Chair of Entrepreneurial Risk at the Swiss Federal Institute of Technology (ETH) in Zurich, is convinced that the statistics of sudden, large price changes on financial markets have certain similarities with the statistics of earthquake tremors and epileptic fits. Per Bak from Denmark, meanwhile, linked crashes in the stock market prices to avalanches in sand piles and congestion in traffic.

Once it was determined that the fluctuations of stock markets is described fairly well by power law distributions, a theory was needed to justify their application. After all, as Stanley admits, statistical observations only provide indications about the relative frequency of extreme events but say nothing about what causes them. So, in order to understand why fluctuations in the stock market and various natural phenomena have similar characteristics, one has to find out what the cause is for the omnipresence of power distributions.

According to Lux, there is one factor in particular that connects physical phenomena to the stock markets: the interplay of innumerable elements cross-linked in a network. Phenomena that boast multiple interactions can nearly always be described by power law distributions. In seismology, for example, this could be small fissures in which energy builds up until it cascades across ever greater scales until an earthquake occurs. This is the conclusion drawn by Sornette who previously did research at the Department of Earth and Space Sciences at UCLA. In epilepsy, the phenomenon would be the interplay of neurons, connected to each other in myriad ways through synapses.

Econo-physicists transferred such ideas to the realm of financial markets and concluded that the distribution of price fluctuations can only be understood by considering the interactions between many investors and the resulting behaviour. As examples of such interactions, Sornette mentions the drive to imitate, herding behaviour, positive feedback, panic reactions, and spontaneous self-organization. But not everybody is satisfied with arguments by analogy. Bruce Mizrach, a financial theorist from Rutgers University, points out that just because different phenomena exhibit similar power distributions by no means implies that they obey the same laws.

But the econo-physicists stick to their guns. Using computer models, they try to verify whether simple behavioural rules between investors (such as "buy this share once its price has dropped by five percent and your colleague is also buying") can, indeed, lead to bursting the bubbles and to market crashes. Of course, this alone would not constitute a proof for the validity of the rules; but it would at least be an indication that they are a useful tool to explain market behaviour.

Blake LeBaron from Brandeis University is a researcher who simulates financial markets with the tools of statistical physics. In his computer programs, many agents interact according to a few, simple rules. Sometimes collective phenomena emerge, such as panic reactions, which should not occur in the "rational world" of classical financial theory. Simulations showed, for example, that investment strategies which were not at all correlated in regular times started to become very much so during times of crisis due to the "irrational" behaviour of interacting investors. Thus, the original intention to avoid risk is turned on its head: portfolios which appeared to be well diversified became very prone to risk when volatility increased. This could well be one of the reasons why crashes are more frequent than would be expected.

Some econo-physicists are not satisfied with just simulating the behaviour of markets; they want to predict extreme events. In an attempt to do so, Sornette founded the Financial Crisis Observatory at the ETH in Zurich. Since physical phenomena obey certain laws, he hopes to develop tools, suitably adapted from the physical sciences, that will help predict future stock market crashes well ahead of time. One of his starting points is the so-called Omori law from geophysics. This law says that an earthquake is typically preceded and followed by aftershocks which are distributed according to a power law. Such characteristic patterns would let Sornette predict the next crash, so he hopes, and give sufficient warning of a looming crisis. Such is his conviction that his methods will work that he put his money where his mouth is and invested a not-insignificant amount of his own money in the stock market.

Chapter 43

Don't Shoot the Messenger

Experts are still busy looking for reasons for the financial markets' recent collapse. Some experts maintain that Wall Street was brought to its knees because of a mathematical formula, ingeniously devised by the Chinese financial expert David X. Li. Li's formula, known as a *Gaussian copula function*, permitted bankers and institutional investors to model complex risks with more ease and accuracy than ever before. What Li offered the financial community was a tool for the assessment of the risks inherent in investing in securities that are correlated in some ways. Largely due to its simplicity, the formula was quickly adopted and eventually embraced by everybody, from investors to banks. Unfortunately, it was too late. The financial community eventually woke up to realize that the formula fails to give correct results in extreme situations.

The financial crisis began when millions of American home-owners whose mortgages had been approved without serious scrutiny of their creditworthiness were no longer able to make their payments. Mortgage lenders were the first to buckle under the stress, soon followed by larger financial institutions and insurance companies. The simultaneous defaults of so many creditors, and the bankruptcies that inevitably followed, had not been foreseen correctly.

It is not as if investors had not been aware of the risks involved when investing in securities whose probabilities to lose all their value correlate. (The correlation measures how one variable moves in line with another; it is important in determining how risky a combination of securities is.) To get a grasp on the danger of simultaneous defaults, Li developed a so-called copula-formula. He published his paper "On Default Correlation: A Copula Function Approach" in the *Journal of Fixed Income* in 2000. (In statistics, a copula is used to couple the behaviour of two or more variables.)

David X. Li had grown up in rural China in the 1960s where he studied economics before moving to Canada on a scholarship. There he studied business administration and actuarial sciences and obtained a Ph.D. in statistics. Since salaries in financial institutions are much more tempting than in academia, Li decided to embark on a career in finance, taking on positions first in Canada, then in the U.S. On Wall Street, Li started borrowing from his work in actuarial science. For example, life insurers calculate the odds that both spouses die within the same year. Taking a cue from such work, Li developed a formula which calculated the probability that several corporations—trading companies, banks, business establishments, real estate holding companies—who hold bundled mortgages, go bankrupt simultaneously. "Suddenly I thought that the problem I was trying to solve was exactly the problem these guys were trying to solve. Default is like the death of a company," said Li a few years later.

Li's formula was simple to use and easy to interpret. It is no wonder that financial managers whose expertise was not mathematics welcomed this apparent solution to default correlation. To them it opened up a world of new possibilities. The formula's main attraction was that it was easily tractable. Andrew Lo, Professor of Finance at MIT, believes that Li's formula was probably the most broadly used tool to model the simultaneous default of several businesses. But one difficulty was usually overlooked. The formula required the input of a parameter which measures to what degree the fortunes of different securities run in parallel. This parameter, called the correlation coefficient, cannot be estimated easily. As indicators, Li used historical data on interest rates that the companies had to pay for their loans. The span between these interest rates and the yields of riskless treasury bills was a proxy for how risky the banks thought the firm was. The data allowed an estimation of the correlation coefficient that was required for the copula formula.

But the use of historical data can be very misleading. This is especially true if the data derives from the decades when real estate prices in the U.S. skyrocketed. Obviously, data from boom years have little relevancy for a looming crisis. For example, it is highly unlikely that in normal times a large number of homeowners default at the same time. But when the housing market started to collapse and the first creditors fell behind with their payments, an avalanche of defaults followed. The basic assumption of Li's formula—namely that the correlation coefficient is a constant parameter—was no longer correct. The probabilities of bankruptcies began to correlate more than the formula had predicted. The riskiness even of diversified portfolios rose.

One asset class after the next was affected. Suddenly, everything was highly correlated and everybody was affected and hurting. Paul Embrechts, a mathematician and financial expert from the ETH in Zurich, had warned in 2001 that the naïve application of simplistic risk models might set off a crisis and even destabilize an entire economy. Traditional risk models just cannot predict exceptional events. The correlation coefficient required for Li's formula, estimated on the basis of historical data, does not suffice when modeling the simultaneous occurrence of several extreme situations.

It would be grossly unfair, however, to fault a formula for the catastrophic consequences of its incorrect use. Lo, the MIT professor, said in Li's defense, that this would be as ridiculous as blaming Newton's laws of motion for the occurrence of fatal accidents. Embrechts points out that mathematicians, in contrast to most financial practitioners, are taught to routinely examine the assumptions on which formulas are based. It is, therefore, not the formula which should be blamed for the crisis, but plain human greed.

Interdisciplinary
matters

Chapter 44

Fascinating Fractals

When Jackson Pollock came onto the art scene in the middle of the previous century, he shocked the world. Critics and connoisseurs were polarized. In the eyes of most of the viewing public, the vast canvasses on which Pollock worked in his famous abstract "drip-and-drop" style were nothing but an arbitrary mess of colours, something that any child could produce. The work of the Dutch artist Piet Mondrian, one of Pollock's contemporaries and supporters, was equally misunderstood by the public. Other than that, the differences between these two artists could not have been greater. While Pollack, a volatile man, partial to alcohol, developed the habit of spontaneously dripping colour on the horizontal canvasses, barely taking a few seconds to do so, Mondrian was a sophisticated intellectual who wrote philosophical essays about his work. He would spend hours deliberating where to position one of his sparse horizontal or vertical lines or coloured rectangles.

In a study published in 2004 in the journal *Chaos and Complexity Letters*, Richard Taylor, a physicist at the University of Oregon, presented the results of a study in which he analyzed the paintings produced by these two startlingly different artists. For Pollock's paintings, Taylor employed a tool that was originally developed for chaos theory, the so-called fractal dimension of an object.

A brush stroke is, of course, a one-dimensional object whereas a canvas is two dimensional. In the 1970s, the French mathematician Benoît Mandelbrot, founder of the so-called theory of fractals, established that in between these simple geometric objects there are complex shapes which have "fractal dimensions" between one and two.

So if a smooth line has dimension one, and a completely filled plane has dimension two, the circumference of a snow flake, which partly fills the plane, lies somewhere in between one and two. In fact, its fractal dimension has been computed to about 1.26. As the complexity and richness of a shape increases, its value moves closer to two. The fact that a natural phenomenon has a fractal dimension— which may not be noticeable to the casual observer—is an indication that its underlying evolution is not haphazard but deterministic.

Fractal objects, from the Latin *fractus* ("broken"), display self-similarity. This means that the same pattern recurs upon finer and finer magnifications. A tree, for example, can be a fractal object since the pattern of the trunk and the branches is repeated in the patterns formed by the branches and the twigs, the twigs and smaller twigs, and so on. The small structures look very much like the whole.

Taylor started his analysis of Pollock's paintings by scanning them into a computer. Covering the scans with a grid of identical squares, he then counted the squares which were occupied by painted patterns and the ones which were left empty. The manner in which the proportion of painted squares to unpainted squares grows as the size of the squares is reduced, and the picture magnified, yields the fractal dimension. Taylor maintains that paintings from amateurs who randomly splatter about paint in an attempt to imitate Pollock's style do not yield consistent estimates of the fractal dimension as the grid becomes finer and finer. In contrast, Pollock's paintings, far from being random, remained fractal over the entire range of grids, from square sizes of 2.5 meters down to one centimetre. This led him to the conclusion that Pollock's paintings are by no means random. The ingenious technique also allowed Taylor to demonstrate that the complexity of Pollock's paintings increased as the artist grew older and refined his technique. The fractal dimension rose from about 1.3 in his early works to approximately 1.8 in his later paintings.

When Pollock once declared "I am Nature" he had only invited the public's ridicule. Yet now, with the help of rigorous mathematical analysis, Taylor was able to prove that Pollock, by sheer intuition, came as close to fulfilling his ambition, as could be expected only from a genius. Amazingly, Pollock had been painting fractals twenty-five years before their discovery by mathematicians and physicists.

And where does this leave Mondrian in whose works there is not the faintest trace of chaos and fractals? He was, after all, notorious for refusing to paint anything other than regular geometric shapes and horizontal and vertical lines. The latter, he believed "exist everywhere and dominate everything." Not even diagonals stood a chance with Mondrian who fiercely believed that they represent disruptive elements disturbing a painting's balance. Are these rules, so passionately formulated, based on any valid aesthetic reason? The answer is a resounding "no." In an experiment, viewers expressed no preference when a Mondrian painting was presented to them in its original orientation or when it was rotated by 45 degrees.

Piet Mondrian not only obsessed about the orientation of his lines, he also worried endlessly about their precise positioning. This too now turns out to be of minor importance to the viewer. Taylor analyzed Mondrian's positioning of horizontal and vertical lines and found out that, statistically, Mondrian was more likely to place them closer to the edges of the canvas than would have been the case had they been placed randomly. Yet in experiments, experts who were shown both genuine paintings by Mondrian and randomly generated paintings "á la Mondrian" again expressed no preference for one or the other type.

Chapter 45

In Dubio (probably) Pro Reo

The principle "in dubio pro reo" (innocent until proven guilty) has been a legal principle for millennia. Based on the legal inference that most people are not criminals, the accused is to be favored. Any guilt must be proven beyond any reasonable doubt for a defendant to be convicted. Bowing to this age-old principle also means, however, that criminals are often set free.

The question is, how high must the probability be for the guilt of an accused to be established beyond reasonable doubt? Ariel Porat and Alon Harel, two law professors at the Center for Rationality and Interactive Decision Theory in Jerusalem, Israel, believe that the mathematical theory of probability can help ascertain whether an accused is guilty of an offence.

Suppose a defendant may only be convicted of a crime if the evidence indicates a probability of 95 percent that the person actually committed the offence. This means that the judiciary agree to keep nineteen possible criminals out of prison in order to avoid the conviction of one possibly innocent person.

Let's take the example of Mr. Smith. He stands accused of having committed two separate offences in two different places and times. Based on the available evidence, the probability that Smith indeed committed these offences is 90 percent in both cases. According to current legal practice, Smith must be acquitted in both cases. But, the probability that Smith is really totally innocent is very small. For both offences there is only a ten percent chance that Smith has not committed them. Harel and Porat now suggest to aggregate the probabilities. The probability that the accused committed neither one nor the other of the two offences can be calculated by multiplying ten percent with itself (0.10×0.10). This amounts to 1 percent (0.01). Hence, Smith committed at least one of the offences with a 99 percent probability.

According to the "Aggregate Probabilities Principle" (APP) proposed by Porat and Harel, Smith should be convicted for at least one of the offences even though the body of evidence, examined separately for each alleged offence, would allow doubts as to his guilt. The traditional practice in our current legal system of acquitting him in both cases due to lack of sufficient evidence is tantamount to 99 possible criminals walking free in order to protect one innocent person from conviction.

There is a flipside, however, to the APP, and it favors the accused. Let us take the example of Miller who is accused of two offences, committed in two different places and times. Let us say the probability that he committed the offences is 95 percent in both cases. A traditional jury would convict Miller for both offences. According to the APP principle, however, the probability that Miller indeed committed both offences is only about 90 percent (0.95×0.95). This would not suffice for convictions in both cases and Miller would be acquitted of one of the offences.

It is possible that judges apply the aggregation principle automatically and unconsciously when they render judgment, for example when they take into consideration prior convictions or unatoned sins. This, however, can lead to quite paradoxical situations.

Let us say that Peter and Paul are each accused of an offence. In both cases the probability of their having committed the crime is 90 percent. In a traditional court, both of the accused would be acquitted due to insufficient evidence. But let us say further that Peter and Paul had been accused some time ago of similar offences.

At that time the evidence brought to court was insufficient to convict Peter—the probability that he had committed the offence was only 90 percent—but Paul was sentenced to prison since the probability that he had committed the offence was 95 percent. According to the principle of aggregation, the judge should now consider the following: the probability of Peter having committed both offences was 90 percent in each case. The aggregated probability of his being quite innocent of both offences is only one percent. (See the calculation above.) Hence, the probability that Peter committed at least one offence is 99 percent and he should be sent to prison.

On the other hand, the probability that Paul committed both offences is only 86 percent (95 percent × 90 percent). Since Paul already completed one prison sentence, the judge should let him go this time. Thus we have the following scenario: in spite of the evidence being identical, the previously convicted Peter is acquitted, while Paul, with a clean record, is incarcerated.

One can think of yet another drawback to the principle of aggregation. Practically everybody would be delinquent if there were sufficient laws that could be broken. Let us assume that there are one hundred traffic rules and that a driver breaks each one of them with an annual probability of three percent. By our reckoning he should definitely have his driver's license revoked after a year since he would be totally innocent of any traffic offense with a probability of less than five percent. (0.97^{100}). To avoid such miscarriages of justice, Harel and Porat would not admit the principle of aggregation for unspecific accusations.

Chapter 46

Once Upon a Time There Was a Mathematical Problem

Mathematics is a discipline with rigorous laws and a reputation for strict precision. Clear-cut definitions, curt theorems, and proofs that are restricted to the most important assertions, are the mathematicians' tool of the trade. No room is left for interpretation. There must be no doubt whatsoever as to what is meant by a declaration and even minute uncertainty about the truth of a statement is anathema to a committed mathematician.

Quite the opposite is true in literature. Vague descriptions, ambiguous allusions and double-entendres are the writer's daily bread and butter. Creative writers have poetic license and are allowed to overstate or downplay. The reader, on his part, responds with a vast spectrum of different reactions. He is free to let his fantasy wander, interpret the text at his whim, make associations. Indeed, he may gain a different understanding of the piece whenever he rereads it, based on his feeling at that particular moment. So can these two forms of creativity, mathematics and *belles lettres*, be reconciled? Or is it a case of math is math, and literature is literature, and never the twain shall meet?

At first glance the latter does seem to be the case. In contrast to natural scientists such as biologists, physicists, or biochemists, mathematicians deal with highly abstract objects that have nothing in common with daily experience. To describe them, mathematicians need to use a special language—not only technical jargon but even a syntax particular to their field. Mathematical papers are so abstract that often not even colleagues working in closely related fields can grasp them. Publications in professional journals have ceased to be vehicles for the dissemination of information. Instead, they simply represent stamps of quality, reserved exclusively for those who are let in on the secret. Readers of such mathematical papers—often no more than a dozen or two spread around the world—are expected to have spent years familiarizing themselves with the subject.

It is not surprising, therefore, that mathematics is considered by the public as something of a secret science. Truth be told, not all mathematicians are unhappy about that. Quite a few among them are quite content in keeping to themselves and going after their research, safely tucked away in their ivory towers. Since mathematical research does not weigh heavily on the public purse, they feel no need to justify their activity. A peaceful yet unsatisfactory coexistence between mathematicians and the general public has become the norm.

Increasingly, however, mathematicians have become aware that the separation of mathematics from general culture does harm to both sides. On the other hand, laypeople have come to recognize that mathematics is an intrinsic part of daily life; they would like to better understand what the subject is all about and how mathematicians go about their trade. Fortunately, authors discovered a new genre of literature in recent years: nonfiction books about mathematics and mathematical novels. With this, a formal end was put to the elitist and isolationist attitude which had prevailed for over two and a half thousand years and had been epitomized by the sign over the entrance of Plato's academy: "Let no one ignorant of geometry enter here."

The beginning of a new era, the time of mathematical storytelling is dawning. The self-imposed shackles which had kept mathematical knowledge a virtual prisoner in the secluded lofts of secretive scientists are being severed, setting the subject free to make its entry into novel territory. Increasingly, mathematical knowledge is being imparted to readers outside the guild, even to people who simply want to be diverted and entertained. Mathematics has come to enjoy a new status, a celebratory status at that. Bestsellers break bookstore records; films such as "A Beautiful Mind" and "Good Will Hunting" have become classics; a TV series such as "Numb3rs" reaches record numbers of viewers; and theatre plays such as "Arcadia" or "Proof" boast sell-out performances.

One writer who has been instrumental in bringing mathematics to the masses is Apostolos Doxiadis whose *Uncle Petros and Goldbach's Conjecture* became an international bestseller. In order to further develop the narrative approach to mathematics he founded the organization "Thales and Friends" which aimed at bridging the gap between the disciplines. One of its first activities was the organization of a conference in Mykonos in the summer of 2005. Under the motto "bridging the chasm between mathematics and human culture," participants explored ways of using a narrative approach to render mathematics palatable also to nonmathematicians.

The fusion of mathematics and narrative is of interest to laymen and professionals alike. Even experts sometimes find relief when professional lingo and the traditional mathematical triad—assumption, proposition, proof—are dispensed with for a while. Persi Diaconis, a world renowned statistician from Stanford University who worked his way through college and graduate school as a magician, admitted that he could only deal with a problem if he also knew its story. Who is concerned with it, how did it arise, what will happen once it is solved? For example, he could only get excited about a certain problem that combined combinatorics, algebra, and function theory after he had cloaked it in a question of how many times a deck of playing cards must be shuffled before their order can be considered sufficiently random. (Answer: seven times.) Similarly, Barry Mazur from Harvard University admitted that he only really comprehended the deeper meaning of a certain question in number theory once he had formulated it in simple, commonly used language in order to explain it to colleagues from other fields.

This, of course, was new territory for the mathematicians-turned-story-tellers gathered in Mykonos. It soon became apparent that rules needed to be established for the new genre. All of a sudden, mathematicians had to grapple with issues which until now had never been their concern. What stylistic devices are admissible? How exact does the wording have to be? May one simplify things for the sake of the reader? How far may one deviate from the rigor to which mathematics has traditionally been wedded? Leo Corry, a historian of science from Tel Aviv University exemplified the dilemma by drawing on an example from another cultural genre—music. Has the movie "Amadeus" helped or hindered the promotion of Mozart's music to the general audience? Did the film, due to its imprecision and many errors, maybe cause irreparable damage? The often heated discussions in Mykonos testified that mathematicians were still far removed from a "unité de doctrine." But there is no denying that in terms of appreciation and entertainment, the public is gaining new insights into this heretofore "forbidden territory" and, vice versa, scientists are gaining a new and increasingly appreciative audience.

Chapter 47

If Only My Ringtone Were Unique

It has happened to everyone. A familiar ringtone alerts to an incoming call. One automatically reaches for the cell phone, only to realize that it was actually the neighbor's phone which rang. Many cell phones have exactly the same ringtone even though everyone thinks that theirs was a truly unique one since it had been downloaded for good money from a specialized website.

There may be relief from such confusion, however. For as little as two dollars anyone can buy his or her very own ringtone from the Wolfram Corporation. The software company promises that every ringtone it provides is, with near certainty, completely different from any other ringtone anywhere in the world.

"Wolfram Tones" is the brainchild of Stephen Wolfram, a distinguished British physicist, who created a sensation a few years ago with his 1200-page book *A New Kind of Science*. Released in 2002 with much fanfare and shameless self-promotion, the book immediately became a bestseller. In it, Wolfram asserted that all natural phenomena are based on processes that can be simulated on computers by cellular automata.

A cellular automaton is a simple computer program; it has nothing in common with cellular telephones but is based on a mathematical theory developed by the German mathematician John von Neumann in the 1940s at the Institute for Advanced Studies in Princeton. Soon after inventing them, von Neumann lost interest in cellular automata and they lay dormant until the 24-year old Wolfram, then at the Institute for Advanced Study on a MacArthur "Genius" Fellowhip, re-discovered them in 1983.

Cellular automata operate on a grid of cells which get painted either black or white, according to the rules of the algorithm. At the outset, the cells in the top row of the grid are randomly coloured. Then the fun begins. The cells in the next row are coloured black or white, according to some rules which depend on the colours of the cells immediately above. The rules are very simple. One of them says, for example, that a cell is to be coloured black if two of the three cells immediately above are black; otherwise the cell is to be coloured white. Another could be that the cell is to be white if the cell immediately above is black and the two cells above and to the right, and above and to the left are white. Once all cells in the second row have been coloured, the operation is repeated in the next row. Then in the next, and the next, and so on.

With such simple rules one would assume, wrongly as it turns out, that only very little of interest can be produced by the algorithm. As it happens, the exact opposite is the case. Depending on which set of rules is applied, the most varied and interesting patterns emerge. Some patterns constantly repeat themselves while others appear to be completely random. Still others exhibit extraordinary richness; even though the pattern appears to have some *order*, it is never predictable. Thanks to Wolfram's research, it is now possible to categorize the complex phenomena which arise as a result of these rules, and to provide the mathematical underpinning to John von Neumann's neglected theory.

In his book, Wolfram argues that all natural phenomena are based on cellular automata. This might actually be true in the case of snowflakes and some sea shells since their patterns indeed remind us of cellular automata. But to claim that the entirety of natural phenomena is based on the repeated application of simple computational transformations is probably a step too far. The fact that the evolution of a phenomenon can be simulated by a computer program is in no way sufficient proof that it actually emerged in this way.

Lately the furore around Wolfram has quieted down. This might have affected his ego but it made little difference to his considerable wealth. His software package "Mathematica"^{TM} which does symbolic calculations and is considered the market leader in the natural and engineering sciences, has sold and is still selling in the millions. The financial success has allowed Wolfram to devote himself to research. One of the happy outcomes of this are the ringtones.

It is not a surprise that Wolfram's ringtones are based on cellular automata. Together with Peter Overmann, a computer scientist, Wolfram developed a process which translates the coloured cells, depending on their position, into musical notes. Astonishingly, the resulting melodies sound pleasing and are far from banal. This can be partly explained by the essence of cellular automata: the generated melodies exhibit sufficient regularity so as not to be totally chaotic, while at the same time displaying enough irregularity to sound interesting.

Wolfram's melodies can be retrieved from the webpage, **tones. wolfram.com**. One first selects the genre, jazz, country, classic, etc. Then, with each click of the mouse, Wolfram's computer program searches through the universe of 10^{27} melodies to produce a new, never-before-heard, thirty-second tune. Some are appealing; others less so. But there is an absolute guarantee that nobody else will ever hit on the same one. For two dollars the customer gets an absolutely unique ringtone that can be adapted to suit one's personal preferences by choosing on what instruments it should be played, and at which tempo. One may even add a drum beat.

Chapter 48

Enforcing Voluntary Cooperation

Many social scientists have asked themselves how altruism and cooperation can evolve among members of a community who act only in their own self interest. Researchers from Harvard University and the University of Vienna carried out a study in which they demonstrated with the help of a mathematical model that cooperation may emerge in a community in which wrongdoers can be penalized. The condition for this to happen is, however, that membership in the community be voluntary.

In the model that the scientists developed, individuals can choose either to receive a secure income or to participate in a risky game. Those who choose to participate can either pay a fee or they can avoid doing so. A profit margin is added to the total income and the proceeds are distributed equally among all players—including the freeloaders. If a sufficient number of donors put up money, all participants benefit. But if too many freeloaders try to profit from the good will of the donors, the latter suffer. In order to avoid this situation, the donors in the model are allowed to impose a penalty on the freeloaders. However, the imposition of the penalty carries a cost. Hence, not every donor will enforce it. In total, there are four strategies: nonparticipants desist from playing; freeloaders participate but do not pay the fee; contributors pay the fee but abstain from imposing penalties; and, finally, enforcers not only pay the fee, but also actively levy a penalty on freeloaders.

The game was simulated on a computer for many consecutive rounds. Players are assigned to the four categories and, in the first few rounds, behave according to their category's strategy. From time to time, they are allowed to modify their behaviour with a certain probability and to adopt the strategy of their more successful colleagues. Also, random changes between categories occur sometimes. The question now is, which strategy prevails as time goes on?

The results of the simulations surprised the researchers: whenever participation in the game was mandatory, with contributors and freeloaders participating, most participants soon ended up as freeloaders. At this point the game had to end because there was no one left to pay fees. This sorry state of affairs did not change when enforcers were added to the game. They could not prevail against the many freeloaders.

However, whenever participation in the game was voluntary—that is, whenever players could opt for a secure income—many freeloaders withdrew from the game in exchange for the secure income. The ones who remained in the game were the contributors, the enforcers, and a few freeloaders. Now, when contributors gained the upper hand over the enforcers, the group could not sustain itself because it would invariably be taken over by the ever-multiplying freeloaders. Hence, the only groups that endured were the ones in which enforcers dominated. They ensured that everybody cooperated.

The paradoxical result is that cooperation can be enforced by penalizing freeloaders, but only if participation in the community is voluntary. This was grasped intuitively by the Nobel Prize-winning Professor of Economics Milton Friedman. Discussing the problems of recruiting sufficient numbers of troops during the Vietnam war, Friedman suggested paying people to join the armed forces. His argument was that he preferred an army of mercenaries to an army of slaves. This reminds us that discipline in the dreaded foreign legion, which legionnaires join of their own free will, is legendary. In the compulsory army, on the other hand, it is often in rather short supply.

Chapter 49

Code or Hoax?

In 1912 the Polish rare book collector Wilfrid Voynich bought a number of medieval manuscripts from a group of Italian Jesuits in Frascati, Italy, for his bookshop in London. Among the dusty piles of folios, he came across a richly illustrated manuscript, written in an unknown script. Its origin can be traced back to Prague at the beginning of the sixteenth century. Emperor Rudolf II of Bohemia, a collector of rare and unusual items, had purchased it for the princely sum of 600 gold ducats. Scientists at the imperial court pored over it for years and studied it in depth. They concluded that Roger Bacon, the thirteenth century Franciscan friar, must have been the author. But to the emperor's great disappointment, nobody could decipher the text. So the manuscript remained unread. Disenchanted, Emperor Rudolf gave it away and the manuscript disappeared for the next four centuries.

After its re-surface at the beginning of the twentieth century, cryptographers, philologists, historians, Vatican archivists, statisticians, and mathematicians again tried to crack the code. The efforts proved fruitless, however. Then, in 2004, a British computer scientist came along and threw all attempts that had hitherto been made to decipher the unknown manuscript overboard. He boldly announced that the manuscript had been penned by a swindler and was bare of any meaning whatsoever. In short, it was nothing but an ancient hoax.

The manuscript, written on good quality vellum, originally consisted of at least 232 pages, of which some have been lost. It measures 15 by 22 centimetres and is about four centimetres thick. While there is a title page and traces of an erased signature can be made out (probably of the owner), neither the title nor the author's name could be identified. Nearly all pages carry illustrations, mostly plants, stars, symbols, and figures of nude females. The main mystery is the text, however. The manuscript remains unread to this day.

The elegant script is reckoned to consist of about three dozen letters and ligations (combinations of letters), none of which bear any relationship to any alphabetic system that has been passed down to us. The text is obviously composed of words that are separated from each other by gaps. Some words are more frequent than others. Based on the illustrations the manuscript appears to be a scientific book consisting of six parts: plants, astronomy, biology, cosmology, pharmaceutics, and recipes.

The first modern examination of the Voynich manuscript was undertaken in the 1920s by William Romaine Newbold, a professor of philosophy at the University of Pennsylvania. Newbold believed that a method based on the rearrangement of letters had been used to encrypt some text. Soon his method of decipherment was shown to be erroneous, however, and the text, soon dubbed "the most mysterious manuscript in the world," remained just that until 1945. Then, a group of cryptologists in Washington, waiting for their demobilization after World War II, tried their hand at deciphering the text. It was not to be. Their efforts to identify a supposed grammar and syntax proved to be in vain, and the same went for their assumption that the manuscript consisted of Latin abbreviations. Equally far off the mark were suggestions that the text represented the stream of words of a schizophrenic, or that it stemmed from the Ukrainian language from which all vowels had been deleted.

Controversy also erupted among those experts who turned toward the illustrations. They vehemently disagreed among themselves as to who was able to identify what and how. One expert, well versed in medieval manuscripts on alchemy, was firmly convinced that the text must have been written before 1460 at the latest. A botanist quickly countered this theory by pointing out that some of the depicted illustrations supposedly reflected New World plants. Hence, he put the manuscript's origin into the early sixteenth century.

Fascinated by the text's aura of mystery, but better equipped than their predecessors, all brought low by the rejection of their carefully conjectured theories, modern scientists set to work. With modern computer technology at their disposal, they were able to perform mathematical analyses of the unknown language. Employing a battery of statistical tests, they measured the frequency of letters, of letter combinations, and of words across wide sections of the texts. They calculated the so-called entropy, a numerical measure of randomness in a string of characters, analyzed the distribution of word lengths and the correlation between words. They employed mathematical methods such as spectral analysis, cluster analysis, and the theory of Markov chains. Despite all this, they did not come any closer to deciphering the text. All they found was that the text reads from left to right, is seemingly divided into two dialects, and employs between twenty-three and thirty individual symbols.

Slightly more hopeful signs came from the studies of the Brazilian mathematician Jorge Stolfi who believed that he was able to distinguish consonants and vowels. He demonstrated that the vast majority of words consisted of between one and three parts which he called prefix, stem, and suffix. But, all in all, nothing had been discovered that had not already been known at Emperor Rudolf's court in Prague.

But still, nobody wanted to believe that a Renaissance scholar might have invented an encryption method that would resist all decipherment methods. Thus, two further alternatives were suggested. Either an unskilled writer had made so many mistakes when encoding the script that a decipherment was quite impossible, or a swindler, aware of Rudolf's fascination with the occult, had played a trick on the emperor, thus cheating him out of a large amount of money. These explanations had their shortcomings, though. The apparent care with which the manuscript had been produced would argue against the thesis of errors in the copying and encryption process. Fraud also seemed unlikely since it would have required an immense effort to produce a manuscript which, though meaningless, displayed so many linguistic structures.

This last claim has, however, been rendered void in the study by Gordon Rugg, a professor of computer science at Keele University in Great Britain. Building on Stolfi's theory of three syllables, he employed a so-called Cardan grille, a device widely used in the sixteenth century to write hidden messages. The approach that Rugg thinks was employed consisted of first filling a table with random syllables. Differing combinations of symbols were entered in varying frequencies into three columns that corresponded to prefix, stem, and suffix. Then Rugg slid the Cardan grille, a sort of template with windows for each of the three syllables, from left to right over the table, transcribing the resulting strings of symbols. The mumbo-jumbo thus generated exhibited a startling similarity to the mess of letters found in the Voynich manuscript.

Since the frequency of words in the resulting text and the manner in which syllables are combined depend on the underlying table, a manuscript thus produced will reflect the statistical properties of the table itself. This explains why previous researchers erroneously believed that the text showed linguistic structure. The two different dialects could be explained by the use of two different tables. Rugg believes that a skilled writer would have required no more than about three months to produce a manuscript of 232 pages. The author of the hoax could have been one Edward Kelley, a hack lawyer, notary, and alchemist at Emperor Rudolf's court who was notorious for his frauds and con games.

Of course, Rugg's explanation falls short of a conclusive proof that the manuscript is a hoax. It does, however, provide a plausible explanation of how the manuscript could have been produced. This will certainly not convince the sceptics, however, and the Voynich manuscript will maintain its aura of mystery for some time to come.

Chapter 50

Crusade Against Sloppy Mathematics

A London-based Canadian mathematician, Douglas Keenan, has made it his mission to lead the battle against the sloppy or malicious use of mathematics. You might think that there isn't much scope for differing opinions in mathematics, but when you are dealing with the interpretation of data, it is entirely possible for divergences of views to arise. Sometimes an incorrect interpretation can be made inadvertently, sometimes even consciously. In the field of climate studies, for example, opposing points of view are often backed up by scientific research that is based on the mathematical analysis of data. Because mathematics gives such work a stamp of credibility, politicians often rely on them. It is therefore all the more important for them to be carried out with care.

After studying mathematics at the University of Waterloo, Keenan worked on Wall Street for a few years but, starting in 1995, devoted himself completely to the forensic study of mathematics. Since then he has been leading, all on his own, a real crusade against shady mathematical machinations. The targets for his often vigorously worded attacks are numerous, and range from the misuse of statistical methods in determining the origin of volcanic ashes to the questionable use of tree rings in evaluating the date of a shipwreck.

Three years ago, the scientific journal *Nature* published a study that used the ripening process of Pinot Noir grapes as an indicator for the warmth of the climate. The official start of the harvest in August is determined by the ripeness of the grapes which, in turn, is determined by the temperature of the summer that has just ended. Since the dates for the beginning of the harvest in Burgundy have been recorded in city archives since 1370, they could conceivably be used as indicators for the way temperatures have developed over the past six centuries. A French research team came up with a model based on this data. The model showed that the summer of 2003 was the hottest in 600 years. The conclusion was clear: Burgundy is warming up.

The work aroused Keenan's suspicion, and he wanted to verify its mathematical foundations. In order to do this however, he needed the raw data—but the authors were not prepared to divulge it. It was only after two requests to *Nature* that they finally handed their documents over. Keenan immediately made a find. The authors had smoothed the data for their study, confused standard errors with standard deviations, used incorrect parameters, and confused daily temperatures with average temperatures. Once all these sources of error are taken into account, the year 2003 does indeed display high temperatures, but not unexpectedly high ones. It is no surprise that the *Nature* editors had not noticed anything, since the data was never put at their disposal, and they never asked to see it either. Had they done so, they would easily have seen through the authors' game. The mere fact that the grape harvest model gave a temperature for 2003 that was 2.4 degrees Celsius above the temperature actually measured by Météo France should have made the editors suspicious.

Keenan's most recent targets are two pieces of work that examine the influence of urbanization on climate change during the period from 1954 to 1983. In order to compare measurements made over time, it is absolutely crucial that the locations of the stations where the measurements are carried out not change throughout the observation period. For example, because a city generates warmth, a measuring station that is moved from the center of the city to its periphery would record lower measurements. On the other hand, the measurements would be more likely to rise if a measuring station was moved from a position upwind from the city, to a position downwind. Even small changes of location, such as, for example, from a field to the asphalt road next to it, lead to deviations. Keenan was, above all, doubtful about the measurements made in China. He did not believe that a scientific study would have been carried out with much care during Mao's Cultural Revolution, when scientists were held in very low esteem.

Once again Keenan found himself running into a brick wall when he asked which stations had been used to make the measurements. "Why should I make the data available to you, when all you want to do is find something wrong with it?" asked one of the authors. But the professor had not reckoned with Keenan's obstinacy. Since the professor was working at a university in England, he was subject to the Freedom of Information Act. This law obliges employees of public institutions to release data. He was thus forced to hand over the list of the Chinese measuring stations to Keenan. And, lo and behold: out of 35 measuring stations, 25 had been subjected to a change of location, sometimes even several changes, which often covered dozens of kilometers. For a further 49 measuring stations, documentation did not even exist.

References

Chapter 2: Arias de Reyna, J. and van de Lune, J. "High Precision Computation of a Constant in the Theory of Trigonometric Series." *Mathematical Computation*, Published electronically, February 9, 2009.

Chapter 3: Kathrin Bringmann and Ken Ono, "Mock Theta Functions", *Proceedings of the National Academy of Sciences*, Vol. **104**, 2007, 3725-3731.

Chapter 4: Thomas Hales, "A proof of the Kepler conjecture", *Annals of Mathematics*, Vol. **162**, Nr. 3, November 2005, 1063-1183.

Chapter 5: A. N. Trahtman, "The Road Coloring Problem", *Israel Journal of Mathematics*, Vol. **172**, 2009, 51-60.

Chapter 6: Régis de la Bretèche, Carl Pomerance and Gérald Tenenbaum, "Products of ratios of consecutive integers", *The Ramanujan Journal*, Vol. **9**, Nr. 102, April 2005, 131-138.

Chapter 10: Manoj Srinivasan, "Optimal speeds for walking and running, and walking on a moving walkway", *Chaos*, Vol. **19**, 2009.

Chapter 7: Ben Green and Terence Tao, "The primes contain arbitrarily long progressions", http://arxiv.org/PS_cache/math/pdf/0404/0404188v6.pdf, 2007.

Chapter 8: Amitabha Tripathi, "A Note on the Postage-Stamp Problem", *Journal of Integer Sequences*, Vol. **9**, Nr. 1, 2006.

Chapter 12: João Gama Oliveira and Albert-László Barabási, "Human dynamics: Darwin and Einstein correspondence patterns", *Nature*, Vol. **437**, 27 October 2005.

Chapter 13: Bill Baritompa, Rainer Löwen, Burkard Polster and Marty Ross, "Mathematical Table Turning Revisited", *Mathematical Intelligencer*, Vol. **29**, 2007, 49-58.

Chapter 18: Ajai Choudhry, "Triads of integers with equal sums of squares, cubes and fourth powers", *Bulletin of the London Mathematical Society*, Vol. **35**, Nr. 6, November 2003, 821-824.

Chapter 20: Daniel K. Biss, "The homotopy type of the matroid Grassmannian", *Annals of Mathematics*, Vol. **158**, Nr. 3, 2003, 929-952.

Daniel K. Biss, "Erratum to 'The homotopy type of the matroid Grassmannian'", *Annals of Mathematics*, Vol. **170**, Number 1, 2009, 493.

Daniel K. Biss, "Oriented matroids, complex manifolds, and a combinatorial model for BU", *Advances in Mathematics*, Vol. **179**, Nr. 2, 2003, 250–290.

Daniel K. Biss, 'Erratum to 'Oriented matroids, complex manifolds, and a combinatorial model for BU'", *Advances in Mathematics*, Vol. **221**, Number 2, 2009, 681.

Chapter 21: Eitan Bachmat, Daniel Berend, Luba Sapir, Steven Skiena, and Natan Stolyarov, "Analysis of Airplane Boarding Times", *Operations Research*, Vol. **57**, Nr. 2, 2009, 499-513.

Chapter 22: Dietrich Braess, "On a paradox of traffic planning", *Transportation Science*, Vol. **39**, 2005, 446-450.

Chapter 23: R. Guimerà, S. Mossa, A. Turtschi, and L. A. N. Amaral, "The worldwide air transportation network: Anomalous centrality, community structure, and cities' global roles", *PNAS*, Vol. **102**, Nr. 22, May 31, 2005, 7794-7799.

Chapter 24: Andrew M. Edwards, Richard A. Phillips, Nicholas W. Watkins, Mervyn P. Freeman, Eugene J. Murphy, Vsevolod Afanasyev, Sergey V. Buldyrev, M. G. E. da Luz, E. P. Raposo, H. Eugene Stanley and Gandhimohan M. Viswanathan, "Revisiting

Lévy flight search patterns of wandering albatrosses, bumblebees and deer", *Nature*, Vol. **449**, 25 October 2007, 1044-1048.

Chapter 25: Roi Cohen Kadosh, Kathrin Cohen Kadosh, Teresa Schuhmann, Amanda Kaas, Rainer Goebel, Avishai Henik, and Alexander T. Sack, "Virtual dyscalculia induced by parietal-lobe TMS impairs automatic magnitude processing", *Current Biology*, Vol. **17**, Nr. 8, 2007, 689-693.

Chapter 26: Rosemary A. Varley, Nicolai J. C. Klessinger, Charles A. J. Romanowski and Michael Siegal, "Agrammatic but numerate", *PNAS*, Vol. **102**, Nr. 9, 2005, 3519-3524.

Chapter 27: Graeme S. Halford, Rosemary Baker, Julie E. Mc-Credden, and John D. Bain, "How Many Variables Can Humans Process?", *Psychological Science*, Vol. **16**, January 2005.

Chapter 29: Daniel Kunkle and Gene Cooperman, "Twenty-Six Moves Suffice for Rubik's Cube." *Proceedings of the International Symposium on Symbolic and Algebraic Computation*, 2007 (ISSAC '07), ACM Press.

Chapter 31: Daniel Schindel, Matthew Rempel and Peter Loly, "Enumerating the bent diagonal squares of Dr. Benjamin Franklin FRS", *Proceedings of the Royal Society A: Physical, Mathematical and Engineering*, Vol. 462, 2006, 2271-2279.

Chapter 32: Andrew J. Tatem, Carlos A. Guerra, Peter M. Atkinson and Simon I. Hay, "Athletics: Momentous sprint at the 2156 Olympics?", *Nature*, Vol. **431**, September 2004, 525.

Chapter 33: Peter D. Sozou and Robert M. Seymour, "Costly but worthless gifts facilitate courtship", *Proceedings of the Royal Society B*, Vol. **272**, Nr. 1575, 22 September 2005, 1877-1884.

Chapter 34: Joseph Beck, "*Combinatorial Games: Tic-Tac-Toe Theory*", Cambridge University Press, England, 2008.

Chapter 35: Joel Spencer and Ioana Dumitriu, "The Liargame over an Arbitrary Channel", *Combinatorica*, Vol. **25**, 2005, 537-559.

Chapter 36: J. Schaeffer, Y. Björnsson, N. Burch, A. Kishimoto, M. Müller, R. Lake, P. Lu and S. Sutphen, "Checkers is Solved", *Science*, Vol. **317**, Nr. 5844, 2007, 1453-1632.

Chapter 37: Robert J. Aumann and M. Maschler, "Game Theoretic Analysis of a Bankruptcy Problem from the *Talmud*," *Journal of Economic Theory*, Vol. **36**, 1985, 195-213.

Chapter 39: Saharon Shelah and Alexander Soifer, "Axiom of choice and chromatic number of the plane", *Journal of Combinatorial Theory A*, Vol. **103**, Nr. 2, August 2003, 387-391.

Saharon Shelah and Alexander Soifer, "Axiom of choice and chromatic number: Examples on the plane", *Journal of Combinatorial Theory A*, Vol. **105**, Nr. 2, January 2004, 359-364.

Alexander Soifer, "Axiom of choice and chromatic number of $\mathbf{R}^n$", *Journal of Combinatorial Theory A*, Vol. **110**, Nr. 1, April 2005, 169-173.

Chapter 40: George G. Szpiro, "*Numbers Rule: The Vexing Mathematics of Democracy, from Plato to the Present*", Princeton University Press, 2010.

Chapter 41: Franz Seitz, Dietrich Stoyan, and Karl-Heinz Tödter, "Coin migration within the euro area", `http://econstor.eu/bitstream/10419/28117/1/609868446.pdf`

Chapter 43: David X. Li, "On Default Correlation: A Copula Function Approach", *Journal of Fixed Income*, Vol. **9**, 2000, 43–54.

Chapter 44: Richard Taylor, "Pollock, Mondrian and Nature: Recent Scientific Investigations", *Chaos and Complexity Letters*, Vol. **1**, 2004.

Chapter 45: Alon Harel and Ariel Porat, "Aggregating Probabilities Across Offences in Criminal Law", Public Law Working Paper Nr. 204, University of Chicago, 2008, `http://papers.ssrn.com/sol3/papers.cfm?abstract_id=1104803`

Chapter 48: Christoph Hauert, Arne Traulsen, Hannelore Brandt, Martin A. Nowak and Karl Sigmund, "Via freedom to coercion: The emergence of costly punishment", *Science*, Vol. **316**, 2007, 1905-1907.